A Recorder's Log Book or Label List of

BRITISH BUTTERFLIES AND MOTHS

J. D. Bradley and D. S. Fletcher

CURWEN BOOKS
1979

Published by Curwen Books

Curwen Press Ltd, Plaistow, London E13
Printed in England
ISBN 0 902068 08 3

A Recorder's Log Book or Label List of British Butterflies and Moths

J. D. Bradley and D. S. Fletcher

A complete list of the Lepidoptera – including subspecies and some biological forms – recorded from the British Isles.

The species are classified under their scientific names, with the English names of the butterflies and larger moths (MACROLEPIDOPTERA) and also those of some smaller moths (MICROLEPIDOPTERA) of economic importance or special interest.

The nomenclature and classification have been brought up to date but essentially follow the revised Lepidoptera part of Kloet & Hincks' *A Check List of British Insects* published by the Royal Entomological Society of London (1972) and the subsequent Addenda & Corrigenda published in the *Entomologist's Gazette* (1974). Where a change in the specific name of a species has since occurred, the synonym is given in italics below the valid name. The log book can thus be conveniently used in conjunction with the Kloet & Hincks check list, particularly when searching the literature for records.

The National Recording Scheme

This log book may be used when recording Lepidoptera for the Biological Records Centre national data bank and mapping schemes. After the records have been noted the log book will be returned to the recorder – *see overleaf.*

Lepidoptera recording scheme

The aim of this scheme is to collect data on Lepidoptera occurring in the British Isles and produce distribution maps, etc.

For MACROLEPIDOPTERA the largest recording unit is the 10 km square of the national grid; for MICROLEPIDOPTERA the largest unit is the vice-county*.

Smaller units may be used for both groups when convenient, e.g. a particular area such as a National Nature Reserve or a named locality on an Ordnance Survey map.

To submit records for inclusion in the scheme, please fill in the relevant sections on the opposite page and send your log book to:

Biological Records Centre,
Institute of Terrestrial Ecology,
Monks Wood Experimental Station,
Huntingdon, Cambs PE17 2LS

* Vice-county maps (published by the Ray Society) are obtainable from: Publications (Sales), British Museum (Natural History), Cromwell Road, London SW7 5BD.

Details required: sections 1 and 2 should be completed, and whichever of sections 3 to 7 are appropriate.

1 Recorder's name and address (BLOCK CAPITALS)

N.B. The log book will be returned to this address

2 Period covered

3 Locality

O.S. map (1:50000/or 1 inch) ☐☐☐

map ref. ☐☐ ☐☐☐☐☐☐

4 County

5 Vice-county name

and no. ☐☐☐

6 1 km square

(from O.S. map) ☐☐ ☐☐☐☐

7 10 km square (from O.S. map) ☐☐ ☐☐

Index to Families

Systematic list of British Lepidoptera

* Species of doubtful British status, including adventitious species and unconfirmed records

† Extinct or believed extinct in the British Isles

ZEUGLOPTERA

MICROPTERIGIDAE

MICROPTERIX Hb.

1 **tunbergella** Fabr. ....
thunbergella auct.

2 **mansuetella** Zell. ....

3 **aureatella** Scop. ....

4 **aruncella** Scop. ....

5 **calthella** Linn. ....

DACNONYPHA

ERIOCRANIIDAE

ERIOCRANIA Zell.

6 **subpurpurella** Haw. ....

7 **chrysolepidella** Zell. ....

8 **unimaculella** Zett. ....

9 **sparrmannella** Bosc ....

10 **salopiella** Stt. ....

11 **haworthi** Bradl. ....

12 **sangii** Wood ....

13 **semipurpurella** Steph. ....

EXOPORIA

HEPIALIDAE

HEPIALUS Fabr.

14 **humuli** Linn.
ssp. **humuli** Linn. ....
Ghost Moth

humuli Linn.
ssp. **thulensis** Newm. (Shetland) ....
Ghost Moth

15 **sylvina** Linn. ....
Orange Swift

16 **hecta** Linn. ....
Gold Swift

17 **lupulinus** Linn. ....
Common Swift

18 **fusconebulosa** DeG.
ssp. **fusconebulosa** DeG. ....
Map-winged Swift

fusconebulosa DeG.
ssp. **shetlandicus** Viette (Shetland) ....
Map-winged Swift

MONOTRYSIA

NEPTICULIDAE

BOHEMANNIA Stt.

19 **quadrimaculella** Boh. ....

ETAINIA Beirne

20 **decentella** H.-S. ....

21 **sericopeza** Zell. ....

22 **sphendamni** Her. ....

ECTOEDEMIA Busck

23 **argyropeza** Zell. ....

24 **turbidella** Zell. ....

25 **intimella** Zell. ....

26 **agrimoniae** Frey ....

27 **spinosella** Joann. ....

28 **angulifasciella** Stt. ....

29 **atricollis** Stt. ....

30 **arcuatella** H.-S. ....

31 **rubivora** Wocke ....

32 **erythrogenella** Joann. ....

33 **bradfordi** Emmet ....

34 **argentipedella** Zell. ....

35 **mediofasciella** Haw. ....

36 **quinquella** Bedell ....

37 **albifasciella** Hein. ....

38 **subbimaculella** Haw. ....

39 **heringi** Toll ....
quercifoliae Toll

40 **pulverosella** Stt. ....

41 **atrifrontella** Stt. ....

FOMORIA Beirne

42 **septembrella** Stt. ....

43 **weaveri** Stt. ....

FEDALMIA Beirne

44 **headleyella** Stt. ....

TRIFURCULA Zell.

45 **griseella** Wolff ....

46 **immundella** Zell. ....

47 **pallidella** Zell. ....

48 **cryptella** Stt. ....

49 **eurema** Tutt ....

STIGMELLA Schr.

50 **aurella** Fabr. ....

51 **fragariella** Heyd. ....
nitens Fol.
gei Wocke

52 **dulcella** Hein. ....

53 **splendidissimella** H.–S. ....

54 **auromarginella** Rich. ....

55 **aeneofasciella** H.–S. ....

56 **dryadella** Hofm. ....

57 **filipendulae** Wocke ....

58 **ulmariae** Wocke ....

59 **poterii** Stt. ....

60 **tengstroemi** Nolck. ....

61 **serella** Stt. ....

*62 **tormentillella** H.–S. ....

63 **marginicolella** Stt. ....

64 **continuella** Stt. ....

65 **speciosa** Frey ....

66 **sorbi** Stt. ....

67 **plagicolella** Stt. ....

68 **salicis** Stt. ....

69 **auritella** Skala ....

70 **obliquella** Hein. ....
vimineticola auct.

71 **repentiella** Wolff ....

72 **myrtillella** Stt. ....

73 **trimaculella** Haw. ....

74 **assimilella** Zell. ....

75 **floslactella** Haw. ....

76 **carpinella** Hein. ....

77 **tityrella** Stt. ....

78 **pomella** Vaugh. ....

79 **perpygmaeella** Doubl. ....
pygmaeella Haw.

80 **ulmivora** Fol. ....

81 **hemargyrella** Koll. ....

82 **paradoxa** Frey ....

83 **atricapitella** Haw. ....

84 **ruficapitella** Haw. ....

85 **suberivora** Stt. ....

86 **roborella** Johan. ....

87 **svenssoni** Johan. ....

88 **samiatella** Zell. ....

89 **basiguttella** Hein. ....

90 **tiliae** Frey ....

91 **minusculella** H.–S. ....

92 **anomalella** Goeze ....
Rose Leaf Miner

93 **centifoliella** Zell. ....

94 **spinosissimae** Waters ....

95 **viscerella** Stt. ....

*96 **ulmiphaga** Preiss. ....

97 **malella** Stt. ....
Apple Pygmy

98 **catharticella** Stt. ....

99 **hybnerella** Hb. ....

100 **oxyacanthella** Stt. ....
aeneella auct.

101 **pyri** Glitz ....

102 **aceris** Frey ....

103 **nylandriella** Tengst. ....
aucupariae Frey

104 **magdalenae** Klim. ....
nylandriella auct.

105 **desperatella** Frey ....

106 **torminalis** Wood ....

107 **regiella** H.–S. ....

108 **crataegella** Klim. ....

109 **prunetorum** Stt. ....

110 **betulicola** Stt. ....

betulicola Stt.
f. **nanivora** Pet. ....

111 **microtheriella** Stt. ....

112 **luteella** Stt. ....

113 **distinguenda** Hein. ....

114 **glutinosae** Stt. ....

115 **alnetella** Stt. ....

116 **lapponica** Wocke ....

117 **confusella** Wood ....

118 **acetosae** Stt. ....

OPOSTEGIDAE

OPOSTEGA Zell.

119 **salaciella** Treit. ....

120 **auritella** Hb. ....

121 **crepusculella** Zell. ....

122 **spatulella** H.–S. ....

TISCHERIIDAE

TISCHERIA Zell.

123 **ekebladella** Bjerk. ....

124 **dodonaea** Stt. ....

125 **marginea** Haw. ....

126 **gaunacella** Dup. ....

127 **angusticollella** Dup. ....

INCURVARIIDAE
INCURVARIINAE

PHYLLOPORIA Hein.

128 **bistrigella** Haw. ....

INCURVARIA Haw.

129 **pectinea** Haw. ....
zinckenii Zell.

130 **masculella** D. & S. ....

LAMPRONIA Steph.

131 **oehlmanniella** Hb. ....

132 **praelatella** D. & S. ....

133 **capitella** Cl. ....
Currant Shoot Borer

*134 **flavimitrella** Hb. ....

135 **luzella** Hb. ....

136 **rubiella** Bjerk. ....
Raspberry Moth

137 **morosa** Zell. ....

138 **fuscatella** Tengst. ....

139 **pubicornis** Haw. ....

NEMATOPOGON Zell.

140 **swammerdamella** Linn. ....

141 **schwarziellus** Zell. ....
panzerella auct.

142 **pilella** D. & S. ....

143 **metaxella** Hb. ....

ADELINAE

NEMOPHORA Ill. & Hoff.

144 **fasciella** Fabr. ....

145 **minimella** D. & S. ....

146 **cupriacella** Hb. ....

147 **metallica** Poda ....
scabiosella Scop.

148 **degeerella** Linn. ....

ADELA Latr.

149 **cuprella** D. & S. ....

150 **reaumurella** Linn. ....

151 **croesella** Scop. ....

152 **rufimitrella** Scop. ....

153 **fibulella** D. & S. ....

HELIOZELIDAE

HELIOZELA H.–S.

154 **sericiella** Haw. ....

*155 **stanneella** F. v. R. ....

156 **resplendella** Stt. ....

157 **hammoniella** Sorh. ....
betulae Stt.

ANTISPILA Hb.

158 **pfeifferella** Hb. ....

159 **petryi** Mart. ....

DITRYSIA

COSSIDAE
ZEUZERINAE

PHRAGMATAECIA Newm.

160 **castaneae** Hb. ....
Reed Leopard

ZEUZERA Latr.

161 **pyrina** Linn. ....
Leopard Moth

COSSINAE

COSSUS Fabr.

162 **cossus** Linn. ....
Goat Moth

ZYGAENIDAE
PROCRIDINAE

ADSCITA Retz.

163 **statices** Linn. ....
The Forester

164 **geryon** Hb.
Cistus Forester ....

165 **globulariae** Hb.
Scarce Forester ....

ZYGAENINAE

ZYGAENA Fabr.

166 **exulans** Hohen.
ssp. **subochracea** White ....
Scotch Burnet

167 **loti** D. & S.
ssp. **scotica** Rowl.–Br. ....
Slender Scotch Burnet

168 **viciae** D. & S.
ssp. **anglica** Reiss ....
New Forest Burnet

viciae D. & S.
ssp. **argyllensis** Trem. (Argyll) ....
New Forest Burnet

169 **filipendulae** Linn.
ssp. **anglicola** Trem. ....
Six-spot Burnet

170 **trifolii** Esp.
ssp. **decreta** Ver. (marshland) ....
Five-spot Burnet

trifolii Esp.
ssp. **palustrella** Ver. (chalkland) ....
Five-spot Burnet

171 **lonicerae** Schev.
ssp. **transferens** Ver. (S. & C. England) ....
Narrow-bordered Five-spot Burnet

lonicerae Schev.
ssp. **latomarginata** Tutt (N.E. England) ....
Narrow-bordered Five-spot Burnet

lonicerae Schev.
ssp. **jocelynae** Trem. (Skye) ....
Narrow-bordered Five-spot Burnet

lonicerae Schev.
ssp. **insularis** Trem. (Ireland) ....
Narrow-bordered Five-spot Burnet

172 **purpuralis** Brünn.
ssp. **segontii** Trem. (N. Wales) ....
Transparent Burnet

purpuralis Brünn.
ssp. **caledonensis** Reiss (Scotland) ....
Transparent Burnet

purpuralis Brünn.
ssp. **sabulosa** Trem. (Ireland) ....
hibernica auct.
Transparent Burnet

LIMACODIDAE

APODA Haw.

173 **limacodes** Hufn. ....
avellana auct.
The Festoon

HETEROGENEA Knoch

174 **asella** D. & S. ....
The Triangle

PSYCHIDAE
SOLENOBIINAE

NARYCIA Steph.

175 **monilifera** Geoff. ....

SOLENOBIA Dup.

176 **triquetrella** Hb. ....

177 **inconspicuella** Stt. ....
Lesser Lichen Case-bearer

*178 **douglasii** Stt. ....

179 **lichenella** Linn. ....
Lichen Case-bearer

TALEPORIINAE

DIPLODOMA Zell.

180 **herminata** Geoff. ....

TALEPORIA Hb.

181 **tubulosa** Retz. ....

BANKESIA Tutt

182 **conspurcatella** Zell. ....

PSYCHINAE

BACOTIA Tutt

183 **sepium** Spey. ....

LUFFIA Tutt

184 **lapidella** Goeze ....

185 **ferchaultella** Steph. ....

PSYCHE Schr.

186 **casta** Pallas ....

187 **crassiorella** Bru. ....

188 **betulina** Zell. ....
eppingella Tutt

EPICHNOPTERIX Hb.

189 **plumella** D. & S. ....

190 **retiella** Newm. ....

OIKETICINAE

ACANTHOPSYCHE Heyl.

191 **atra** Linn. ....

PACHYTHELIA Westw.

192 **villosella** Ochs. ....

CANEPHORA Hb.

*193 **unicolor** Hufn. ....

THYRIDOPTERYX Steph.

*194 **ephemeraeformis** Haw. ....

STERROPTERIX Hb.

195 **fusca** Haw. ....

TINEIDAE
SCARDIINAE

MOROPHAGA H.–S.

196 **choragella** D. & S. ....
boleti Fabr.

EUPLOCAMINAE

EUPLOCAMUS Latr.

*197 **anthracinalis** Scop. ....

DRYADAULINAE

DRYADAULA Meyr.

198 **pactolia** Meyr. ....

TEICHOBIINAE

PSYCHOIDES Bru.

199 **verhuella** Bru. ....

200 **filicivora** Meyr. ....

MEESSIINAE

LICHENOVORA Pet.

201 **nigripunctella** Haw. ....

MEESSIA Hofm.

202 **richardsoni** Wals. ....

INFURCITINEA Spul.

203 **argentimaculella** Stt. ....

204 **albicomella** H.–S. ....

ISCHNOSCIA Meyr.

205 **borreonella** Mill. ....

CELESTICA Meyr.

206 **angustipennis** H.–S. ....

MYRMECOZELINAE

MYRMECOZELA Zell.

207 **ochraceella** Tengst. ....

ATELIOTUM Zell.

*208 **insularis** Rebel ....

SETOMORPHINAE

SETOMORPHA Zell.

*209 **rutella** Zell. ....
Tropical Tobacco Moth

LINDERA Blanch.

*210 **tessellatella** Blanch. ....

NEMAPOGONINAE

HAPLOTINEA Diak. & Hint.

211 **ditella** Pier. & Metc. ....

212 **insectella** Fabr. ....

CEPHIMALLOTA Bru.

213 **angusticostella** Zell. ....

CEPHITINEA Zag.

*214 **colongella** Zag. ....

NEMAPOGON Schr.

215 **granella** Linn. ....
Corn Moth

216 **cloacella** Haw.
Cork Moth

217 **wolffiella** Karsh. & Niel.
albipunctella Haw.

218 **personella** Pier. & Metc.

219 **ruricolella** Stt.

220 **clematella** Fabr.
arcella auct.

221 **picarella** Cl.

ARCHINEMAPOGON Zag.

222 **laterella** Thunb.

NEMAXERA Zag.

223 **betulinella** Fabr.
corticella Curt.

TRIAXOMERA Zag.

224 **parasitella** Hb.

225 **fulvimitrella** Sodof.

TRIAXOMASIA Zag.

226 **caprimulgella** Stt.

TINEINAE

MONOPIS Hb.

227 **rusticella** Hb.
Skin Moth

228 **weaverella** Scott

229 **ferruginella** Hb.

230 **crocicapitella** Clem.

231 **imella** Hb.

232 **monachella** Hb.

233 **fenestratella** Heyd.

TRICHOPHAGA Rag.

234 **tapetzella** Linn.
Tapestry Moth

*235 **mormopis** Meyr.
percna Corb. & Tams

TINEOLA H.-S.

236 **bisselliella** Hum.
Common Clothes Moth

NIDITINEA Pet.

237 **fuscipunctella** Haw. ....
Brown-dotted Clothes Moth

238 **piercella** Bent. ....

TINEA Linn.

239 **columbariella** Wocke ....

240 **pellionella** Linn. ....
Case-bearing Clothes Moth

241 **lanella** Pier. & Metc. ....

242 **metonella** Pier. & Metc. ....

243 **turicensis** Müll.–Rutz ....

244 **flavescentella** Haw. ....

245 **pallescentella** Stt. ....
Large Pale Clothes Moth

246 **semifulvella** Haw. ....

247 **trinotella** Thunb. ....

247a **fictrix** Meyr. ....

TENAGA Clem.

*248 **pomiliella** Clem. ....

CERATOPHAGA Pet.

*249 **orientalis** Stt. ....

*250 **haidarabadi** Zag. ....

OCHSENHEIMERIIDAE

OCHSENHEIMERIA Hb.

251 **mediopectinellus** Haw. ....

252 **bisontella** Lien. & Zell. ....

253 **vacculella** F. v. R. ....

LYONETIIDAE
CEMIOSTOMINAE

LEUCOPTERA Hb.

254 **laburnella** Stt. ....
Laburnum Leaf Miner

255 **wailesella** Stt. ....

256 **spartifoliella** Hb. ....

257 **orobi** Stt. ....

258 **lathyrifoliella** Stt. ....

259 **lotella** Stt. ....

260 **scitella** Zell. ....
Pear Leaf Blister Moth

PARALEUCOPTERA Heinr.

261 **sinuella** Reutti ....

LYONETIINAE

LYONETIA Hb.

262 **prunifoliella** Hb. ....

263 **clerkella** Linn. ....
Apple Leaf Miner

BEDELLIINAE

BEDELLIA Stt.

264 **somnulentella** Zell. ....

BUCCULATRIGINAE

BUCCULATRIX Zell.

265 **cristatella** Zell. ....

266 **nigricomella** Zell. ....

267 **maritima** Stt. ....

268 **capreella** Krog. ....

*269 **artemisiella** H.–S. ....

270 **frangulella** Goeze ....

271 **albedinella** Zell. ....

272 **cidarella** Zell. ....

273 **thoracella** Thunb. ....

274 **ulmella** Zell. ....

275 **crataegi** Zell. ....

276 **demaryella** Dup. ....

HIEROXESTIDAE

OINOPHILA Steph.

277 **v-flava** Haw. ....
Yellow V Moth

OPOGONA Zell.

*278 **sacchari** Bojer ....

*279 **antistacta** Meyr ....

GRACILLARIIDAE
GRACILLARIINAE

CALOPTILIA Hb.

280 **cuculipennella** Hb. ....
281 **populetorum** Zell. ....
282 **elongella** Linn. ....
283 **betulicola** Her. ....
284 **rufipennella** Hb. ....
285 **azaleella** Brants ....
Azalea Leaf Miner
286 **alchimiella** Scop. ....
287 **robustella** Jäckh ....
288 **stigmatella** Fabr. ....
289 **falconipennella** Hb. ....
290 **semifascia** Haw. ....
291 **hemidactylella** D. & S. ....
292 **leucapennella** Steph. ....
293 **syringella** Fabr. ....

ASPILAPTERYX Spul.

294 **tringipennella** Zell. ....

CALYBITES Hb.

295 **pyrenaeella** Chret. ....
296 **phasianipennella** Hb. ....
297 **auroguttella** Steph.

MICRURAPTERYX Spul.

*298 **kollariella** Zell. ....

PARECTOPA Clem.

299 **ononidis** Zell. ....

PARORNIX Spul.

300 **loganella** Stt. ....
301 **betulae** Stt. ....
302 **fagivora** Frey ....
303 **anglicella** Stt. ....
304 **devoniella** Stt. ....
305 **scoticella** Stt. ....
306 **alpicola** Wocke ....
307 **leucostola** Pel.–Clint. ....

308 **finitimella** Zell. ....

309 **torquillella** Zell. ....

CALLISTO Steph.

310 **denticulella** Thunb. ....

ACROCERCOPS Wall.

311 **imperialella** Zell. ....

*312 **hofmanniella** Schl. ....

313 **brongniardella** Fabr. ....

LEUCOSPILAPTERYX Spul.

314 **omissella** Stt. ....

LITHOCOLLETINAE

PHYLLONORYCTER Hb.

315 **harrisella** Linn. ....

316 **roboris** Zell. ....

317 **heegeriella** Zell. ....

318 **tenerella** Joann. ....

319 **saportella** Dup. ....

320 **quercifoliella** Zell. ....

321 **messaniella** Zell. ....

322 **muelleriella** Zell. ....

323 **oxyacanthae** Frey ....

324 **sorbi** Frey ....

325 **mespilella** Hb. ....

326 **blancardella** Fabr. ....

327 **cydoniella** D. & S. ....

328 **junoniella** Zell. ....

329 **pomonella** Zell. ....

330 **cerasicolella** H.–S. ....

331 **lantanella** Schr. ....

332 **corylifoliella** Hb. ....

333 **viminiella** Sirc. ....

334 **viminetorum** Stt. ....

335 **salicicolella** Sirc. ....

336 **dubitella** H.–S. ....

337 **spinolella** Dup. ....

338 **cavella** Zell. ....

339 **ulicicolella** Stt. ....

340 **scopariella** Zell. ....

341 **maestingella** Müll. ....

342 **coryli** Nic. ....
Nut Leaf Blister Moth

343 **quinnata** Geoff. ....

344 **strigulatella** Zell. ....
rajella auct.
alnifoliella auct.

345 **rajella** Linn. ....
alnifoliella Hb.

346 **distentella** Zell. ....

347 **anderidae** Fletch. ....

348 **quinqueguttella** Stt. ....

349 **nigrescentella** Logan ....

350 **insignitella** Zell. ....

351 **lautella** Zell. ....

352 **schreberella** Fabr. ....

353 **ulmifoliella** Hb. ....

354 **emberizaepenella** Bouché ....

355 **scabiosella** Dougl. ....

356 **tristrigella** Haw. ....

357 **stettinensis** Nic. ....

358 **froelichiella** Zell. ....

359 **nicellii** Stt. ....

360 **kleemannella** Fabr. ....

361 **trifasciella** Haw. ....

362 **sylvella** Haw. ....

363 **platanoidella** Joann. ....

364 **geniculella** Rag. ....

365 **comparella** Dup. ....

366 **sagitella** Bjerk. ....

PHYLLOCNISTIDAE

PHYLLOCNISTIS Zell.

367 **saligna** Zell. ....

368 **unipunctella** Steph. ....

369 **xenia** Her. ....

SESIIDAE
SESIINAE

SESIA Fabr.

370 **apiformis** Cl. ....
Hornet Moth

371 **bembeciformis** Hb. ....
Lunar Hornet Moth

PARANTHRENINAE

PARANTHRENE Hb.

372 **tabaniformis** Rott. ....
Dusky Clearwing

SYNANTHEDON Hb.

373 **tipuliformis** Cl. ....
salmachus L.
Currant Clearwing

374 **vespiformis** Linn. ....
Yellow-legged Clearwing

375 **spheciformis** D. & S. ....
White-barred Clearwing

376 **scoliaeformis** Borkh. ....
Welsh Clearwing

377 **flaviventris** Stgdr ....
Sallow Clearwing

378 **anthraciniformis** Esp. ....
Orange-tailed Clearwing

379 **myopaeformis** Borkh. ....
Red-belted Clearwing

380 **formicaeformis** Esp. ....
Red-tipped Clearwing

381 **culiciformis** Linn. ....
Large Red-belted Clearwing

BEMBECIA Hb.

382 **scopigera** Scop. ....
Six-belted Clearwing

383 **muscaeformis** Esp. ....
Thrift Clearwing

384 **chrysidiformis** Esp. ....
Fiery Clearwing

CHOREUTIDAE

ANTHOPHILA Haw.

385 **fabriciana** Linn. ....

TEBENNA Billb.

*386 **bjerkandrella** Thunb. ....

CHOREUTIS Hb.

387 **sehestediana** Fabr. ....
punctosa Haw.

388 **myllerana** Fabr. ....

EUTROMULA Fröl.

389 **pariana** Cl. ....
Apple Leaf Skeletonizer

390 **diana** Hb. ....

GLYPHIPTERIGIDAE

GLYPHIPTERIX Hb.

391 **simpliciella** Steph. ....
Cocksfoot Moth

392 **schoenicolella** Boyd ....

393 **minorella** Snell. ....
equitella auct.

394 **forsterella** Fabr. ....

395 **haworthana** Steph. ....

396 **fuscoviridella** Haw. ....

397 **thrasonella** Scop. ....

DOUGLASIIDAE

TINAGMA Zell.

398 **ocnerostomella** Stt. ....

399 **balteolella** F. v. R. ....

HELIODINIDAE

HELIODINES Stt.

400 **roesella** Linn. ....

YPONOMEUTIDAE
ARGYRESTHIINAE

ARGYRESTHIA Hb.

401 **laevigatella** H.–S. ....

*402 **illuminatella** Zell. ....

403 **glabratella** Zell. ....

404 **praecocella** Zell. ....

405 **arceuthina** Zell. ....

406 **abdominalis** Zell. ....

407 **dilectella** Zell. ....

408 **aurulentella** Stt. ....

409 **ivella** Haw. ....

410 **brockeella** Hb. ....

411 **goedartella** Linn. ....

412 **pygmaeella** Hb. ....

413 **sorbiella** Treit. ....

414 **arcella** Fabr. ....
cornella auct.

415 **retinella** Zell. ....

416 **glaucinella** Zell. ....

417 **mendica** Haw. ....

418 **conjugella** Zell. ....
Apple Fruit Moth

419 **semifusca** Haw. ....

420 **pruniella** Cl. ....
Cherry Fruit Moth

421 **curvella** Linn. ....

422 **albistria** Haw. ....

423 **semitestacella** Curt. ....

YPONOMEUTINAE

YPONOMEUTA Latr.

424 **evonymella** Linn. ....
Bird-cherry Ermine

425 **padella** Linn. ....
Orchard Ermine

426 **malinellus** Zell. ....
Apple Ermine

427 **cagnagella** Hb. ....
Spindle Ermine

428 **rorrella** Hb. ....
Willow Ermine

429 **irrorella** Hb. ....

430 **plumbella** D. & S. ....

431 **vigintipunctata** Retz. ....

EUHYPONOMEUTA Toll

432 **stannella** Thunb. ....

KESSLERIA Now.

433 **fasciapennella** Stt. ....

434 **saxifragae** Stt. ....

ZELLERIA Stt.

435 **hepariella** Stt. ....

PSEUDOSWAMMERDAMIA Friese

436 **combinella** Hb. ....

SWAMMERDAMIA Hb.

437 **caesiella** Hb. ....

438 **pyrella** Vill. ....

439 **compunctella** H.–S. ....

PARASWAMMERDAMIA Friese

440 **spiniella** Hb. ....

441 **lutarea** Haw. ....

CEDESTIS Zell.

442 **gysseleniella** Zell. ....

443 **subfasciella** Steph. ....

OCNEROSTOMA Zell.

444 **piniariella** Zell. ....

445 **friesei** Svens. ....

ROESLERSTAMMIA Zell.

*446 **pronubella** D. & S. ....

447 **erxlebella** Fabr. ....

ATEMELIA H.–S.

448 **torquatella** Zell. ....

PRAYS Hb.

449 **fraxinella** Bjerk. ....
Ash Bud Moth

SCYTHROPIA Hb.

450 **crataegella** Linn. ....
Hawthorn Moth

PLUTELLINAE

YPSOLOPHA Latr.

451 **mucronella** Scop. ....

452 **nemorella** Linn. ....

453 **dentella** Fabr. ....
Honeysuckle Moth

454 **asperella** Linn. ....

455 **scabrella** Linn. ....

456 **horridella** Treit. ....

457 **lucella** Fabr. ....

458 **alpella** D. & S. ....

459 **sylvella** Linn. ....

460 **parenthesella** Linn. ....

461 **ustella** Cl. ....

462 **sequella** Cl. ....

463 **vittella** Linn. ....

PLUTELLA Schr.

464 **xylostella** Linn. ....
Diamond-back Moth

465 **porrectella** Linn. ....

RHIGOGNOSTIS Stdgr

466 **senilella** Zett. ....

467 **annulatella** Curt. ....

468 **incarnatella** Steud. ....

EIDOPHASIA Steph.

469 **messingiella** F. v. R. ....

ORTHOTAELIINAE

ORTHOTAELIA Steph.

470 **sparganella** Thunb. ....

ACROLEPIINAE

DIGITIVALVA Gaedike

471 **perlepidella** Stt. ....

472 **pulicariae** Klim. ....

ACROLEPIOPSIS Gaedike

473 **assectella** Zell. ....
Leek Moth

474 **betulella** Curt. ....

475 **marcidella** Curt. ....

ACROLEPIA Curt.

476 **pygmeana** Haw. ....

EPERMENIIDAE

PHAULERNIS Meyr.

477 **dentella** Zell. ....

478 **fulviguttella** Zell. ....

CATAPLECTICA Wals.

479 **farreni** Wals. ....

480 **profugella** Stt. ....

EPERMENIA Hb.

481 **illigerella** Hb. ....

482 **insecurella** Stt. ....

483 **chaerophyllella** Goeze ....

484 **aequidentellus** Hofm. ....

SCHRECKENSTEINIIDAE

SCHRECKENSTEINIA Hb.

485 **festaliella** Hb. ....

COLEOPHORIDAE

AUGASMA H.-S.

486 **aeratella** Zell. ....

METRIOTES H.-S.

487 **lutarea** Haw. ....

GONIODOMA Zell.

488 **limoniella** Stt. ....

COLEOPHORA Hb.

489 **leucapennella** Hb. ....

490 **lutipennella** Zell. ....

491 **gryphipennella** Hb. ....

492 **flavipennella** Dup. ....

493 **serratella** Linn. ....

494 **coracipennella** Hb. ....

495 **cerasivorella** Pack. ....
Apple & Plum Case-bearer

496 **milvipennis** Zell. ....

497 **badiipennella** Dup. ....

498 **alnifoliae** Bar. ....

499 **limosipennella** Dup. ....

500 **hydrolapathella** Her. ....

501 **siccifolia** Stt. ....

502 **trigeminella** Fuchs ....

503 **fuscocuprella** H.–S. ....

504 **viminetella** Zell. ....

505 **idaeella** Hofm. ....

506 **vitisella** Gregs. ....

507 **glitzella** Hofm. ....

508 **arctostaphyli** Meder ....

509 **violacea** Ström ....
hornigi Toll

510 **juncicolella** Stt. ....

511 **orbitella** Zell. ....

512 **binderella** Koll. ....

513 **potentillae** Elisha ....

514 **ahenella** Hein. ....

515 **albitarsella** Zell. ....

516 **trifolii** Curt. ....
Large Clover Case-bearer

517 **frischella** Linn. ....
alcyonipennella auct.
Small Clover Case-bearer

518 **spissicornis** Haw. ....

519 **deauratella** Lien. & Zell. ....

520 **fuscicornis** Zell. ....

521 **conyzae** Zell. ....

522 **lineolea** Haw. ....

523 **hemerobiella** Scop. ....

524 **lithargyrinella** Zell. ....

525 **solitariella** Zell. ....

526 **laricella** Hb. ....
Larch Case-bearer

527 **wockeella** Zell. ....

528 **chalcogrammella** Zell. ....

529 **tricolor** Wals. ....

530 **lixella** Zell. ....

531 **ochrea** Haw. ....

532 **albidella** D. & S. ....

533 **anatipennella** Hb. ....
Pistol Case-bearer

534 **currucipennella** Zell. ....
535 **ardeaepennella** Scott ....
536 **ibipennella** Zell. ....
537 **palliatella** Zinck. ....
538 **vibicella** Hb. ....
539 **conspicuella** Zell. ....
540 **vibicigerella** Zell. ....
541 **pyrrhulipennella** Zell. ....
542 **serpylletorum** Her. ....
543 **vulnerariae** Zell. ....
544 **albicosta** Haw. ....
545 **saturatella** Stt. ....
546 **genistae** Stt. ....
547 **discordella** Zell. ....
548 **niveicostella** Zell. ....
549 **onosmella** Brahm ....
550 **silenella** H.–S. ....
551 **otitae** Zell. ....
552 **lassella** Stdgr ....
553 **striatipennella** Nyl. ....
554 **inulae** Wocke ....
555 **troglodytella** Dup. ....
556 **trochilella** Dup. ....
557 **machinella** Bradl. ....
558 **ramosella** Zell. ....
559 **peribenanderi** Toll ....
560 **paripennella** Zell. ....
561 **therinella** Tengst. ....
562 **asteris** Mühl. ....
563 **argentula** Steph. ....
564 **virgaureae** Stt. ....
565 **benanderi** Kanerva ....
annulatella auct.
566 **sternipennella** Zett. ....
flavaginella Lien. & Zell.
moeniacella Stt.
567 **adspersella** Ben. ....
568 **versurella** Zell. ....
569 **squamosella** Stt. ....

570 **pappiferella** Hofm. ....

*571 **granulatella** Zell. ....

572 **vestianella** Linn. ....
laripennella Zett.
annulatella Nyl.

573 **atriplicis** Meyr. ....

574 **suaedivora** Meyr. ....
moeniacella auct.

575 **salinella** Stt. ....

576 **artemisiella** Scott ....

577 **artemisicolella** Bru. ....

578 **murinipennella** Dup. ....

579 **antennariella** H.–S. ....

580 **sylvaticella** Wood ....

581 **taeniipennella** H.–S. ....

582 **glaucicolella** Wood ....

583 **tamesis** Waters ....

584 **alticolella** Zell. ....

585 **maritimella** Newm. ....

586 **adjunctella** Hodgk. ....

587 **caespititiella** Zell. ....

588 **salicorniae** Wocke ....

589 **clypeiferella** Hofm. ....

ELACHISTIDAE

PERITTIA Stt.

590 **obscurepunctella** Stt. ....

MENDESIA Joann.

*591 **farinella** Thunb. ....

STEPHENSIA Stt.

592 **brunnichella** Linn. ....

ELACHISTA Treit.

593 **regificella** Sirc. ....

594 **gleichenella** Fabr. ....

595 **biatomella** Stt. ....

596 **poae** Stt. ....

597 **atricomella** Stt. ....

598 **kilmunella** Stt. ....

599 **alpinella** Stt. ....
600 **luticomella** Zell. ....
601 **albifrontella** Hb. ....
602 **apicipunctella** Stt. ....
603 **subnigrella** Dougl. ....
604 **orstadii** Palm ....
605 **pomerana** Frey ....
606 **humilis** Zell. ....
607 **pulchella** Haw. ....
608 **rufocinerea** Haw. ....
609 **cerusella** Hb. ....
610 **argentella** Cl. ....
611 **triatomea** Haw. ....
612 **collitella** Dup. ....
613 **subocellea** Steph. ....
614 **triseriatella** Stt. ....
615 **dispunctella** Dup. ....
616 **bedellella** Sirc. ....
617 **megerlella** Hb. ....
618 **cingillella** H.–S. ....
619 **unifasciella** Haw. ....
620 **gangabella** Zell. ....
621 **subalbidella** Schläg. ....
622 **revinctella** Zell. ....
adscitella Stt.
623 **bisulcella** Dup. ....

BISELACHISTA
Tr.–O. & Niel.

624 **trapeziella** Stt. ....
625 **cinereopunctella** Haw. ....
626 **serricornis** Stt. ....
627 **scirpi** Stt. ....
628 **eleochariella** Stt. ....
629 **utonella** Frey ....
630 **albidella** Nyl. ....

COSMIOTES Clem.

631 **freyerella** Hb. ....
632 **consortella** Stt. ....
633 **stabilella** Stt. ....

OECOPHORIDAE
OECOPHORINAE

SCHIFFERMUELLERIA Hb.

634 **grandis** Desv. ....

635 **subaquilea** Stt. ....

636 **similella** Hb. ....

637 **tinctella** Hb. ....

CHAMBERSIA Riley

638 **augustella** Hb. ....

BISIGNA Toll

639 **procerella** D. & S. ....

BATIA Steph.

640 **lunaris** Haw. ....

641 **lambdella** Don. ....

642 **unitella** Hb. ....

DAFA Hodges

†643 **formosella** D. & S. ....

BORKHAUSENIA Hb.

644 **fuscescens** Haw. ....

645 **minutella** Linn. ....

TELECHRYSIS Toll

646 **tripuncta** Haw. ....

HOFMANNOPHILA Spul.

647 **pseudospretella** Stt. ....
Brown House-moth

ENDROSIS Hb.

648 **sarcitrella** Linn. ....
White-shouldered House-moth

ESPERIA Hb.

649 **sulphurella** Fabr. ....

650 **oliviella** Fabr. ....

OECOPHORA Latr.

651 **bractella** Linn. ....

ALABONIA Hb.

652 **geoffrella** Linn. ....

APLOTA Steph.

653 **palpella** Haw. ....

PLEUROTA Hb.

654 **bicostella** Cl. ....

*655 **aristella** Linn. ....

PAROCYSTOLA Turn.

656 **acroxantha** Meyr. ....

HYPERCALLIA Steph.

657 **citrinalis** Scop. ....

CARCINA Hb.

658 **quercana** Fabr. ....

AMPHISBATIS Zell.

659 **incongruella** Stt. ....

PSEUDATEMELIA Rebel

660 **josephinae** Toll ....

661 **flavifrontella** D. & S. ....

662 **subochreella** Doubl. ....

CHIMABACHINAE

DIURNEA Haw.

663 **fagella** D. & S. ....

664 **phryganella** Hb. ....

CHEIMOPHILA Hb.

665 **salicella** Hb. ....

DEPRESSARIINAE

SEMIOSCOPIS Hb.

666 **avellanella** Hb. ....

667 **steinkellneriana** D. & S. ....

ENICOSTOMA Steph.

668 **lobella** D. & S. ....

DEPRESSARIA Haw.

669 **discipunctella** H.–S. ....

670 **daucella** D. & S. ....
nervosa auct.

671 **ultimella** Stt. ....

672 **pastinacella** Dup. ....
heracliana auct.
Parsnip Moth

673 **pimpinellae** Zell. ....

674 **badiella** Hb. ....

*675 **brunneella** Rag. ....

676 **pulcherrimella** Stt. ....

677 **douglasella** Stt. ....

678 **weirella** Stt. ....

*679 **emeritella** Stt. ....

680 **albipunctella** Hb. ....

681 **olerella** Zell. ....

682 **chaerophylli** Zell. ....

683 **depressana** Fabr. ....
depressella Fabr.

684 **silesiaca** Hein. ....

LEVIPALPUS Hann.

685 **hepatariella** Lien. & Zell. ....

EXAERETIA Stt.

686 **ciniflonella** Lien. & Zell. ....

687 **allisella** Stt. ....

AGONOPTERIX Hb.

688 **heracliana** Linn. ....
applana Fabr.

689 **ciliella** Stt. ....

690 **cnicella** Treit. ....

691 **purpurea** Haw. ....

692 **subpropinquella** Stt. ....

693 **putridella** D. & S. ....

694 **nanatella** Stt. ....

695 **alstroemeriana** Cl. ....

696 **propinquella** Treit. ....

697 **arenella** D. & S. ....

698 **liturella** D. & S. ....

699 **bipunctosa** Curt. ....

700 **pallorella** Zell. ....

701 **ocellana** Fabr. ....

702 **assimilella** Treit. ....

703 **pulverella** Hb. ....

704 **scopariella** Hein. ....

705 **ulicetella** Stt. ....

706 **nervosa** Haw. ....
costosa Haw.

*707 **prostratella** Const. ....

708 **carduella** Hb. ....

709 **liturosa** Haw. ....

710 **conterminella** Zell. ....

711 **curvipunctosa** Haw. ....

*712 **astrantiae** Hein. ....

713 **angelicella** Hb. ....

714 **yeatiana** Fabr. ....

715 **capreolella** Zell. ....

716 **rotundella** Dougl. ....

ETHMIIDAE

ETHMIA Hb.

717 **terminella** Fletch. ....

718 **dodecea** Haw. ....

719 **funerella** Fabr. ....

720 **bipunctella** Fabr. ....

*721 **pusiella** Linn. ....

*722 **pyrausta** Pallas ....

GELECHIIDAE
GELECHIINAE

METZNERIA Zell.

723 **littorella** Dougl. ....

724 **lappella** Linn. ....

725 **aestivella** Zell. ....

726 **metzneriella** Stt. ....

727 **neuropterella** Zell. ....

PALTODORA Meyr.

728 **cytisella** Curt. ....

ISOPHRICTIS Meyr.

729 **striatella** D. & S. ....

APODIA Hein.

730 **bifractella** Dup. ....

EULAMPROTES Bradl.

731 **atrella** D. & S. ....

732 **unicolorella** Dup. ....

733 **wilkella** Linn. ....

ARGOLAMPROTES Ben.

734 **micella** D. & S. ....

MONOCHROA Hein.

735 **tenebrella** Hb. ....

736 **lucidella** Steph. ....

737 **palustrella** Dougl. ....

738 **tetragonella** Stt. ....

739 **conspersella** H.–S. ....

740 **hornigi** Stdgr ....

741 **suffusella** Dougl. ....

742 **lutulentella** Zell. ....

743 **elongella** Hein. ....

744 **arundinetella** Stt. ....

745 **divisella** Dougl. ....

CHRYSOESTHIA Hb.

746 **drurella** Fabr. ....
hermannella auct.

747 **sexguttella** Thunb. ....

PTOCHEUUSA Hein.

748 **paupella** Zell. ....

SITOTROGA Hein.

749 **cerealella** Ol. ....
Angoumois Grain Moth

PSAMATHOCRITA Meyr.

750 **osseella** Stt. ....

750a **argentella** Pier. & Metc. ....

ARISTOTELIA Hb.

751 **subdecurtella** Stt. ....

752 **ericinella** Zell. ....

753 **brizella** Treit. ....

XYSTOPHORA Wocke

754 **pulveratella** H.–S. ....

STENOLECHIA Meyr.

755 **gemmella** Linn. ....

PARACHRONISTIS Meyr.

756 **albiceps** Zell. ....

RECURVARIA Haw.

757 **nanella** D. & S. ....

758 **leucatella** Cl. ....

PULICALVARIA Freem.

759 **piceaella** Kearf. ....

EXOTELEIA Wall.

760 **dodecella** Linn. ....

RHYNCHOPACHA Stdgr

761 **tetrapunctella** Thunb. ....

762 **mouffetella** Linn. ....

XENOLECHIA Meyr.

763 **aethiops** Humph. & Westw. ....

PSEUDOTELPHUSA Janse

764 **scalella** Scop. ....

TELEIODES Sattl.

765 **vulgella** Hb. ....

766 **scriptella** Hb. ....

767 **decorella** Haw. ....

768 **notatella** Hb. ....

769 **wagae** Now. ....

770 **proximella** Hb. ....

771 **alburnella** Zell. ....

772 **fugitivella** Zell. ....

773 **paripunctella** Thunb. ....

774 **luculella** Hb. ....

775 **sequax** Haw. ....

TELEIOPSIS Sattl.

776 **diffinis** Haw. ....

BRYOTROPHA Hein.

777 **basaltinella** Zell. ....

778 **umbrosella** Zell. ....

779 **affinis** Haw. ....
780 **similis** Stt. ....
781 **mundella** Dougl. ....
782 **senectella** Zell. ....
783 **borella** Dougl. ....
784 **galbanella** Zell. ....
*785 **figulella** Stdgr ....
786 **desertella** Dougl. ....
787 **terrella** D. & S. ....
788 **politella** Stt. ....
789 **domestica** Haw. ....

CHIONODES Hb.

790 **fumatella** Dougl. ....
791 **distinctella** Zell. ....

MIRIFICARMA Gozm.

792 **mulinella** Zell. ....
793 **lentiginosella** Zell. ....

LITA Treit.

794 **virgella** Thunb. ....
795 **solutella** Zell. ....

AROGA Busck

796 **velocella** Zell. ....

NEOFACULTA Gozm.

797 **ericetella** Geyer ....

NEOFRISERIA Sattl.

798 **peliella** Treit. ....
799 **singula** Stdgr ....

GELECHIA Hb.

800 **rhombella** D. & S. ....
801 **scotinella** H.–S. ....
*802 **sabinella** Zell. ....
802a **sororculella** Hb. ....
803 **muscosella** Zell. ....
804 **cuneatella** Dougl. ....
805 **hippophaella** Schr. ....
806 **nigra** Haw. ....

807 **turpella** D. & S. ....

PLATYEDRA Meyr.

808 **subcinerea** Haw. ....

PEXICOPIA Common

809 **malvella** Hb. ....
Hollyhock Seed Moth

SCROBIPALPA Janse

810 **suaedella** Rich. ....

811 **samadensis** Pfaff.
ssp. **plantaginella** Stt. ....

812 **instabilella** Dougl. ....

813 **salinella** Zell. ....

814 **ocellatella** Boyd ....
Beet Moth

815 **nitentella** Fuchs ....

816 **obsoletella** F. v. R. ....

817 **clintoni** Pov. ....

818 **atriplicella** F. v. R. ....

819 **costella** Humph. & Westw. ....

820 **artemisiella** Treit. ....
Thyme Moth

821 **murinella** H.–S. ....

822 **acuminatella** Sirc. ....

SCROBIPALPULA Pov.

823 **psilella** H.–S. ....

GNORIMOSCHEMA Busck

824 **streliciella** H.–S. ....

PHTHORIMAEA Meyr.

*825 **operculella** Zell. ....
Potato Tuber Moth

CARYOCOLUM Greg. & Pov.

826 **vicinella** Dougl. ....

827 **alsinella** Zell.
ssp. **semidecandrella** Threl. ....

828 **viscariella** Stt. ....

829 **marmoreum** Haw. ....

830 **fraternella** Dougl. ....

831 **proximum** Haw. ....

832 **blandella** Dougl. ....

833 **junctella** Dougl. ....

834 **tricolorella** Haw. ....

835 **blandulella** Tutt ....

836 **kroesmanniella** H.–S. ....
huebneri auct.

837 **huebneri** Haw. ....
knaggsiella Stt.

NOTHRIS Hb.

838 **verbascella** Hb. ....

839 **congressariella** Bru. ....

REUTTIA Hofm.

840 **subocellea** Steph. ....

SOPHRONIA Hb.

841 **semicostella** Hb. ....

842 **humerella** D. & S. ....

APROAEREMA Durr.

843 **anthyllidella** Hb. ....

SYNCOPACMA Meyr.

844 **larseniella** Gozm. ....

845 **sangiella** Stt. ....

846 **vinella** Bankes ....

847 **taeniolella** Zell. ....

848 **albipaepella** H.–S. ....

849 **cinctella** Cl. ....

*850 **polychromella** Rebel ....

ACANTHOPHILA Hein.

851 **alacella** Zell. ....

ANACAMPSIS Curt.

852 **temerella** Lien. & Zell. ....

853 **populella** Cl. ....

populella Cl.
f. **fuscatella** Bent. ....

854 **blattariella** Hb. ....

ACOMPSIA Hb.

855 **cinerella** Cl. ....

ANARSIA Zell.

856 **spartiella** Schr. ....

*857 **lineatella** Zell. ....
Peach Twig Borer

HYPATIMA Hb.

858 **rhomboidella** Linn. ....

PSORICOPTERA Stt.

859 **gibbosella** Zell. ....

MESOPHLEPS Hb.

860 **silacella** Hb. ....

TELEPHILA Meyr.

861 **schmidtiellus** Heyd. ....

DICHOMERIS Hb.

862 **marginella** Fabr. ....
Juniper Webber

863 **juniperella** Linn. ....

864 **ustalella** Fabr. ....

865 **fasciella** Hb. ....

BRACHMIA Hb.

866 **blandella** Fabr. ....
gerronella Zell.

867 **inornatella** Dougl. ....

868 **rufescens** Haw. ....

869 **lutatella** H.–S. ....

SYMMOCINAE

OEGOCONIA Stt.

870 **quadripuncta** Haw. ....

871 **deauratella** H.–S. ....

SYMMOCA Hb.

*872 **signatella** H.–S. ....

BLASTOBASIDAE

BLASTOBASIS Zell.

873 **lignea** Wals. ....

874 **decolorella** Woll. ....

*875 **phycidella** Zell. ....

AUXIMOBASIS Wals.

*876 **normalis** Meyr. ....

STATHMOPODIDAE

STATHMOPODA H.–S.

877 **pedella** Linn. ....

MOMPHIDAE
BATRACHEDRINAE

BATRACHEDRA H.–S.

878 **praeangusta** Haw. ....

879 **pinicolella** Zell. ....

MOMPHINAE

MOMPHA Hb.

880 **langiella** Hb. ....
epilobiella Roem.

881 **terminella** Humph. & Westw. ....

882 **locupletella** D. & S. ....

883 **raschkiella** Zell. ....

884 **miscella** D. & S. ....

885 **conturbatella** Hb. ....

886 **ochraceella** Curt. ....

887 **lacteella** Steph. ....

888 **propinquella** Stt. ....

889 **divisella** H.–S. ....

890 **subdivisella** Bradl. ....

891 **nodicolella** Fuchs ....

892 **subbistrigella** Haw. ....

893 **epilobiella** D. & S. ....
fulvescens Haw.

COSMOPTERIGINAE

COSMOPTERIX Hb.

894 **zieglerella** Hb. ....

895 **schmidiella** Frey ....

896 **orichalcea** Stt. ....
drurella auct.
druryella auct.

897 **lienigiella** Lien. & Zell. ....

LIMNAECIA Stt.

898 **phragmitella** Stt. ....

PANCALIA Steph.

899 **leuwenhoekella** Linn. ....

900 **latreillella** Curt. ....

EUCLEMENSIA Grote

†901 **woodiella** Curt. ....

BLASTODACNINAE

GLYPHIPTERYX Curt.

902 **lathamella** Fletch. ....

903 **linneella** Cl. ....

SPULERIA Hofm.

904 **flavicaput** Haw. ....

BLASTODACNA Wocke

905 **hellerella** Dup. ....

906 **atra** Haw. ....
Apple Pith Moth

DYSTEBENNA Spul.

907 **stephensi** Stt. ....

CHRYSOPELEIINAE

SORHAGENIA Spul.

908 **rhamniella** Zell. ....

909 **lophyrella** Dougl. ....

910 **janiszewskae** Riedl ....

SCYTHRIDIDAE

SCYTHRIS Hb.

911 **grandipennis** Haw. ....

912 **fuscoaenea** Haw. ....

913 **fallacella** Schläg. ....

914 **fletcherella** Meyr. ....

915 **picaepennis** Haw. ....

916 **siccella** Zell. ....

917 **empetrella** Karsh. & Niel. ....
variella Steph.

918 **limbella** Fabr. ....
quadriguttella Thunb.

919 **cicadella** Zell. ....

920 **potentillae** Zell. ....

COCHYLIDAE

HYSTEROSIA Steph.

921 **inopiana** Haw. ....

922 **schreibersiana** Fröl. ....

923 **sodaliana** Haw. ....

HYSTEROPHORA Obraz.

924 **maculosana** Haw. ....

PHTHEOCHROA Steph.

925 **rugosana** Hb. ....

PHALONIDIA Le March.

926 **manniana** F. v. R. ....

927 **minimana** Carad. ....

928 **permixtana** D. & S. ....

929 **vectisana** Humph. & Westw. ....

930 **alismana** Rag. ....

931 **luridana** Gregs. ....

932 **affinitana** Dougl. ....

933 **gilvicomana** Zell. ....

934 **curvistrigana** Stt. ....

STENODES Guen.

935 **alternana** Steph. ....

936 **straminea** Haw. ....

AGAPETA Hb.

937 **hamana** Linn. ....

938 **zoegana** Linn. ....

AETHES Billb.

939 **tesserana** D. & S. ....

940 **rutilana** Hb. ....

941 **hartmanniana** Cl. ....

942 **piercei** Obraz. ....

943 **margarotana** Dup. ....

944 **williana** Brahm ....

945 **cnicana** Westw. ....

946 **rubigana** Treit. ....

947 **smeathmanniana** Fabr. ....

948 **margaritana** Haw. ....

949 **dilucidana** Steph. ....

950 **francillana** Fabr. ....

951 **beatricella** Wals. ....

COMMOPHILA Hb.

952 **aeneana** Hb. ....

EUGNOSTA Hb.

*953 **lathoniana** Hb. ....

EUPOECILIA Steph.

954 **angustana** Hb.
ssp. **angustana** Hb. ....

angustana Hb.
ssp. **angustana** Hb.
f. **fasciella** Don. (heathland & moorland)..........

angustana Hb.
ssp. **thuleana** Vaugh. (Shetland)..........

955 **ambiguella** Hb.
Vine Moth ....

COCHYLIDIA Obraz.

956 **implicitana** Wocke ....

957 **heydeniana** H.–S. ....

958 **subroseana** Haw. ....

959 **rupicola** Curt. ....

FALSEUNCARIA
Obraz. & Swats.

960 **ruficiliana** Haw. ....

961 **degreyana** McLach. ....

COCHYLIS Treit.

962 **roseana** Haw. ....

963 **flaviciliana** Westw. ....

964 **dubitana** Hb. ....

965 **hybridella** Hb. ....

966 **atricapitana** Steph. ....

967 **pallidana** Zell. ....

968 **nana** Haw. ....

TORTRICIDAE
TORTRICINAE

PANDEMIS Hb.

969 **corylana** Fabr. ....
Chequered Fruit-tree Tortrix

970 **cerasana** Hb. ....
Barred Fruit-tree Tortrix

971 **cinnamomeana** Treit. ....

972 **heparana** D. & S. ....
Dark Fruit-tree Tortrix

973 **dumetana** Treit. ....

ARGYROTAENIA Steph.

974 **pulchellana** Haw. ....

HOMONA Walk.

*975 **menciana** Walk. ....
Camellia Tortrix

ARCHIPS Hb.

976 **oporana** Linn. ....

977 **podana** Scop. ....
Large Fruit-tree Tortrix

978 **betulana** Hb. ....

979 **crataegana** Hb. ....
Brown Oak Tortrix

980 **xylosteana** Linn. ....
Variegated Golden Tortrix

981 **rosana** Linn. ....
Rose Tortrix

CHORISTONEURA Led.

982 **diversana** Hb. ....

983 **hebenstreitella** Müll. ....

984 **lafauryana** Rag. ....

CACOECIMORPHA Obraz.

985 **pronubana** Hb. ....
Carnation Tortrix

SYNDEMIS Hb.

986 **musculana** Hb.
ssp. **musculana** Hb. ....

musculana Hb.
ssp. **musculinana** Kenn. (Shetland) ....

PTYCHOLOMOIDES Obraz.

987 **aeriferanus** H.–S. ....

APHELIA Hb.

988 **viburnana** D. & S. ....
Bilberry Tortrix

989 **paleana** Hb. ....
Timothy Tortrix

990 **unitana** Hb. ....

CLEPSIS Guen.

991 **senecionana** Hb. ....

992 **rurinana** Linn. ....

993 **spectrana** Treit. ....
Cyclamen Tortrix

994 **consimilana** Hb. ....

*995 **trileucana** Doubl. ....

*996 **melaleucanus** Walk. ....

EPICHORISTODES Diak.

*997 **acerbella** Walk. ....
African Carnation Tortrix

EPIPHYAS Turn.

998 **postvittana** Walk. ....
Light Brown Apple Moth

ADOXOPHYES Meyr.

999 **orana** F. v. R. ....
Summer Fruit Tortrix

PTYCHOLOMA Steph.

1000 **lecheana** Linn. ....

LOZOTAENIODES Obraz.

1001 **formosanus** Geyer ....

LOZOTAENIA Steph.

1002 **forsterana** Fabr. ....

*1003 **subocellana** Steph. ....

PARAMESIA Steph.

1004 **gnomana** Cl. ....

PERICLEPSIS Bradl.

1005 **cinctana** D. & S. ....

EPAGOGE Hb.

1006 **grotiana** Fabr. ....

CAPUA Steph.

1007 **vulgana** Fröl. ....

PHILEDONE Hb.

1008 **gerningana** D. & S. ....

PHILEDONIDES Obraz.

1009 **lunana** Thunb. ....

DITULA Steph.

1010 **angustiorana** Haw. ....
Red-barred Tortrix

PSEUDARGYROTOZA Obraz.

1011 **conwagana** Fabr. ....

SPARGANOTHIS Hb.

1012 **pilleriana** D. & S. ....

OLINDIA Guen.

1013 **schumacherana** Fabr. ....

ISOTRIAS Meyr.

1014 **rectifasciana** Haw. ....

EULIA Hb.

1015 **ministrana** Linn. ....

CNEPHASIA Curt.

1016 **longana** Haw. ....

*1017 **gueneana** Dup. ....

1018 **communana** H.-S. ....

1019 **conspersana** Dougl. ....

1020 **stephensiana** Doubl. ....
Grey Tortrix

stephensiana Doubl.
f. **octomaculana** Steph. ....

1021 **interjectana** Haw. ....
Flax Tortrix

1022 **pasiuana** Hb. ....

1023 **genitalana** Pier. & Metc. ....

1024 **incertana** Treit. ....
Light Grey Tortrix

TORTRICODES Guen.

1025 **alternella** D. & S. ....

EXAPATE Hb.

1026 **congelatella** Cl. ....

NEOSPHALEROPTERA Real

1027 **nubilana** Hb. ....

EANA Billb.

1028 **argentana** Cl. ....

1029 **osseana** Scop. ....

1030 **incanana** Steph. ....

1031 **penziana** Thunb. & Beck.
ssp. **penziana** Thunb. & Beck.
f. **bellana** Curt. (N. England, inland) ....

penziana Thunb. & Beck.
ssp. **colquhounana** Barr. (British Is., coastal) ....

ALEIMMA Hb.

1032 **loeflingiana** Linn. ....

TORTRIX Linn.

1033 **viridana** Linn. ....
Green Oak Tortrix

SPATALISTIS Meyr.

1034 **bifasciana** Hb. ....

CROESIA Hb.

1035 **bergmanniana** Linn. ....

1036 **forsskaleana** Linn. ....

1037 **holmiana** Linn. ....

ACLERIS Hb.

1038 **laterana** Fabr. ....
latifasciana Haw.

1037 **comariana** Lien. & Zell. ....
Strawberry Tortrix

1040 **caledoniana** Steph. ....

1041 **sparsana** D. & S. ....

1042 **rhombana** D. & S. ....
Rhomboid Tortrix

1043 **aspersana** Hb. ....

1044 **ferrugana** D. & S. ....

1045 **notana** Don. ....
tripunctana Hb.

1046 **shepherdana** Steph. ....

1047 **schalleriana** Linn. ....

1048 **variegana** D. & S. ....
Garden Rose Tortrix

1049 **permutana** Dup. ....

1050 **boscana** Fabr. ....

1051 **logiana** Cl. ....

1052 **umbrana** Hb. ....

1053 **hastiana** Linn. ....

1054 **cristana** D. & S. ....

1055 **hyemana** Haw. ....

1056 **lipsiana** D. & S. ....

1057 **rufana** D. & S. ....

1058 **lorquiniana** Dup. ....

1059 **abietana** Hb. ....

1060 **maccana** Treit. ....

1061 **literana** Linn. ....

1062 **emargana** Fabr. ....

OLETHREUTINAE

CELYPHA Hb.

1063 **striana** D. & S. ....

1064 **rosaceana** Schläg. ....

1065 **rufana** Scop. ....

1066 **woodiana** Barr. ....

OLETHREUTES Hb.

1067 **cespitana** Hb. ....

1068 **rivulana** Scop. ....

1069 **aurofasciana** Haw. ....

1070 **mygindiana** D. & S. ....

1071 **arbutella** Linn. ....

1072 **metallicana** Hb. ....

1073 **schulziana** Fabr. ....

1074 **palustrana** Lien. & Zell. ....

1075 **olivana** Treit. ....

1076 **lacunana** D. & S. ....

1077 **obsoletana** Zett. ....

1078 **doubledayana** Barr. ....

1079 **bifasciana** Haw. ....

1080 **arcuella** Cl. ....

PRISTEROGNATHA Obraz.

1081 **penthinana** Guen. ....

HEDYA Hb.

1082 **pruniana** Hb. ....
Plum Tortrix

1083 **nubiferana** Haw. ....
Marbled Orchard Tortrix

1084 **ochroleucana** Fröl. ....

1085 **atropunctana** Zett. ....

1086 **salicella** Linn. ....

ORTHOTAENIA Steph.

1087 **undulana** D. & S. ....

PSEUDOSCIAPHILA Obraz.

1088 **branderiana** Linn. ....

APOTOMIS Hb.

1089 **semifasciana** Haw. ....

1090 **infida** Heinr. ....

1091 **lineana** D. & S. ....

1092 **turbidana** Hb. ....

1093 **betuletana** Haw. ....

1094 **capreana** Hb. ....

1095 **sororculana** Zett. ....

1096 **sauciana** Fröl.
ssp. **sauciana** Fröl. (England, Wales, Ireland)..........

sauciana Fröl.
ssp. **grevillana** Curt. (Scotland) ....

ENDOTHENIA Steph.

1097 **gentianaeana** Hb. ....

1098 **oblongana** Haw. ....

1099 **marginana** Haw. ....

1100 **pullana** Haw. ....
fuligana auct.

1101 **ustulana** Haw. ....

1102 **nigricostana** Haw. ....

1103 **ericetana** Humph. & Westw. ....

1104 **quadrimaculana** Haw. ....

LOBESIA Guen.

1105 **occidentis** Falk. ....

1106 **reliquana** Hb. ....

*1107 **botrana** D. & S.
European Vine Moth ....

1108 **abscisana** Doubl. ....

1109 **littoralis** Humph. & Westw. ....

BACTRA Steph.

1110 **furfurana** Haw. ....

1111 **lancealana** Hb. ....

1112 **robustana** Christ. ....

EUDEMIS Hb.

1113 **profundana** D. & S. ....

1114 **porphyrana** Hb. ....

ANCYLIS Hb.

1115 **achatana** D. & S. ....

1116 **comptana** Fröl. ....

1117 **unguicella** Linn. ....

1118 **uncella** D. & S. ....

1119 **geminana** Don. ....

geminana Don.
f. **diminutana** Haw. ....

geminana Don.
f. **subarcuana** Dougl. ....

1120 **mitterbacheriana** D. & S. ....

1121 **upupana** Treit. ....

1122 **obtusana** Haw. ....

1123 **laetana** Fabr. ....

1124 **tineana** Hb. ....

1125 **unculana** Haw. ....

1126 **badiana** D. & S. ....

1127 **paludana** Barr. ....

1128 **myrtillana** Treit. ....

1129 **apicella** D. & S. ....

EPINOTIA Hb.

1130 **pygmaeana** Hb. ....

1131 **subsequana** Haw. ....

1132 **subocellana** Don. ....

1133 **bilunana** Haw. ....

1134 **ramella** Linn. ....

1135 **demarniana** F. v. R. ....

1136 **immundana** F. v. R. ....

1137 **tetraquetrana** Haw. ....

1138 **nisella** Cl. ....

nisella Haw.
f. **cinereana** Haw. ....

1139 **tenerana** D. & S. ....
Nut Bud Moth

1140 **nigricana** H.–S. ....

1141 **nemorivaga** Tengst. ....

1142 **tedella** Cl. ....

1143 **fraternana** Haw. ....

1144 **signatana** Dougl. ....

1145 **nanana** Treit. ....

1146 **rubiginosana** H.–S. ....

1147 **cruciana** Linn. ....
Willow Tortrix

1148 **mercuriana** Fröl. ....

1149 **crenana** Hb. ....

1150 **abbreviana** Fabr. ....
trimaculana Don.

1151 **stroemiana** Fabr. ....

1152 **maculana** Fabr. ....

1153 **sordidana** Hb. ....

1154 **caprana** Fabr. ....

1155 **brunnichana** Linn. ....

1156 **solandriana** Linn. ....

CROCIDOSEMA Zell.

1157 **plebejana** Zell. ....

RHOPOBOTA Led.

1158 **ustomaculana** Curt. ....

1159 **naevana** Hb. ....
unipunctana Haw.
Holly Tortrix

ACROCLITA Led.

1160 **subsequana** H.–S. ....

GRISELDA Heinr.

1161 **stagnana** D. & S. ....

1162 **myrtillana** Humph. & Westw. ....

ZEIRAPHERA Treit.

1163 **ratzeburgiana** Ratz. ....

1164 **rufimitrana** H.–S. ....

1165 **isertana** Fabr. ....

1166 **diniana** Guen. ....
Larch Tortrix

GYPSONOMA Meyr.

1167 **aceriana** Dup. ....

1168 **sociana** Haw. ....

1169 **dealbana** Fröl. ....

1170 **oppressana** Treit. ....

1171 **minutana** Hb. ....

1172 **nitidulana** Lien. & Zell. ....

GIBBERIFERA Obraz.

1173 **simplana** F. v. R. ....

EPIBLEMA Hb.

1174 **cynosbatella** Linn. ....

1175 **uddmanniana** Linn. ....
Bramble Shoot Moth

1176 **trimaculana** Haw. ....

1177 **rosaecolana** Doubl. ....

1178 **roborana** D. & S. ....

1179 **incarnatana** Hb. ....

1180 **tetragonana** Steph. ....

1181 **grandaevana** Lien & Zell. ....

1182 **turbidana** Treit. ....

1183 **foenella** Linn. ....

1184 **scutulana** D. & S. ....

scutulana D. & S.
f. **cirsiana** Zell. ....

1185 **cnicicolana** Zell. ....

1186 **farfarae** Fletch. ....

1187 **costipunctana** Haw. ....

PELOCHRISTA Led.

1188 **caecimaculana** Hb. ....

ERIOPSELA Guen.

1189 **quadrana** Hb. ....

EUCOSMA Hb.

1190 **aspidiscana** Hb. ....

1191 **heringiana** Jäckh ....

1192 **conterminana** H.–S. ....

1193 **tripoliana** Barr. ....

1194 **aemulana** Schläg. ....

1195 **maritima** Humph. & Westw. ....

*1196 **metzneriana** Treit. ....

1197 **campoliliana** D. & S. ....

1198 **pauperana** Dup. ....

1199 **pupillana** Cl. ....

1200 **hohenwartiana** D. & S. ....

hohenwartiana D. & S.
f. **fulvana** Steph. ....

1201 **cana** Haw. ....

1202 **obumbratana** Lien. & Zell. ....

FOVEIFERA Obraz.

*1203 **hastana** Hb. ....

THIODIA Hb.

1204 **citrana** Hb. ....

SPILONOTA Steph.

1205 **ocellana** D. & S. ....
Bud Moth

ocellana D. & S.
f. **laricana** Hein. ....

CLAVIGESTA Obraz.

1206 **sylvestrana** Curt. ....

1207 **purdeyi** Durr. ....
Pine Leaf-mining Moth

BLASTESTHIA Obraz.

1208 **posticana** Zett. ....

1209 **turionella** Linn. ....
Pine Bud Moth

RHYACIONIA Hb.

1210 **buoliana** D. & S. ....
Pine Shoot Moth

1211 **pinicolana** Doubl. ....

1212 **pinivorana** Lien. & Zell. ....
Spotted Shoot Moth

1213 **duplana** Hb.
ssp. **logaea** Durr. ....
Elgin Shoot Moth

PETROVA Heinr.

1214 **resinella** Linn. ....
Pine Resin-gall Moth

CRYPTOPHLEBIA Wals.

*1215 **leucotreta** Meyr. ....
False Codling Moth

ENARMONIA Hb.

1216 **formosana** Scop. ....
Cherry-bark Moth

EUCOSMOMORPHA Obraz.

1217 **albersana** Hb. ....

SELANIA Steph.

1218 **leplastriana** Curt. ....

LATHRONYMPHA Meyr.

1219 **strigana** Fabr. ....

COLLICULARIA Obraz.

1220 **microgrammana** Guen. ....

STROPHEDRA H.–S.

1221 **weirana** Dougl. ....

1222 **nitidana** Fabr. ....

PAMMENE Hb.

1223 **splendidulana** Guen. ....

1224 **luedersiana** Sorh. ....

1225 **obscurana** Steph. ....

*1226 **agnotana** Rebel ....

1227 **inquilana** Fletch. ....

1228 **argyrana** Hb. ....

1229 **albuginana** Guen. ....

*1230 **suspectana** Lien. & Zell. ....

1231 **spiniana** Dup. ....

1232 **populana** Fabr. ....

1233 **aurantiana** Stdgr ....

1234 **regiana** Zell. ....

1235 **trauniana** D. & S. ....

1236 **fasciana** Linn. ....

fasciana Linn.
f. **herrichiana** Hein. ....

1237 **germmana** Hb. ....

1238 **ochsenheimeriana** Lien. & Zell. ....

1239 **rhediella** Cl. ....
Fruitlet Mining Tortrix

CYDIA Hb.

1240 **caecana** Schläg. ....

1241 **compositella** Fabr. ....

1242 **internana** Guen. ....

1243 **pallifrontana** Lien. & Zell. ....

1244 **gemmiferana** Treit. ....

1245 **janthinana** Dup. ....

1246 **tenebrosana** Dup. ....

1247 **funebrana** Treit. ....
Plum Fruit Moth

*1248 **molesta** Busck ....
Oriental Fruit Moth

1249 **prunivorana** Rag. ....

1250 **lathyrana** Hb. ....

1251 **jungiella** Cl. ....

1252 **lunulana** D. & S. ....
dorsana auct.

1253 **orobana** Treit. ....

1254 **strobilella** Linn. ....

1255 **succedana** D. & S. ....

1256 **servillana** Dup. ....

1257 **nigricana** Fabr. ....
Pea Moth

1258 **millenniana** Adamcz. ....
deciduana Steuer

1259 **fagiglandana** Zell. ....

1260 **splendana** Hb. ....

1261 **pomonella** Linn. ....
Codling Moth

*1262 **amplana** Hb. ....

*1263 **inquinatana** Hb. ....

1264 **leguminana** Lien. & Zell. ....

1265 **cognatana** Barr. ....

1266 **pactolana** Zell. ....

1267 **cosmophorana** Treit. ....

1268 **coniferana** Ratz. ....

1269 **conicolana** Heyl. ....

†1270 **corollana** Hb. ....

1271 **gallicana** Guen. ....

1272 **aurana** Fabr. ....

DICHRORAMPHA Guen.

1273 **petiverella** Linn. ....

1274 **alpinana** Treit. ....

1275 **flavidorsana** Knaggs ....

1276 **plumbagana** Treit. ....

1277 **senectana** Guen. ....

1278 **sequana** Hb. ....

1279 **acuminatana** Lien. & Zell. ....

1280 **consortana** Steph. ....

1281 **simpliciana** Haw. ....

1282 **sylvicolana** Hein. ....

1283 **montanana** Dup. ....

1284 **gueneeana** Obraz. ....

1285 **plumbana** Scop. ....

1286 **sedatana** Busck ....

1287 **aeratana** Pier. & Metc. ....

ALUCITIDAE

ALUCITA Linn.

1288 **hexadactyla** Linn. ....
Twenty-plume Moth

PYRALIDAE
CRAMBINAE

EUCHROMIUS Guen.

*1289 **ocellea** Haw. ....

CHILO Zinck.

1290 **phragmitella** Hb. ....

ACIGONA Hb.

*1291 **cicatricella** Hb. ....

CALAMOTROPHA Zell.

1292 **paludella** Hb. ....

CHRYSOTEUCHIA Hb.

1293 **culmella** Linn. ....
hortuella Hb.

CRAMBUS Fabr.

1294 **pascuella** Linn. ....

*1295 **leucoschalis** Hamps. ....

1296 **silvella** Hb. ....

1297 **uliginosellus** Zell. ....

1298 **ericella** Hb. ....

1299 **hamella** Thunb. ....

1300 **pratella** Linn. ....

1301 **nemorella** Hb. ....
pratellus auct.

1302 **perlella** Scop. ....

AGRIPHILA Hb.

1303 **selasella** Hb. ....

1304 **straminella** D. & S. ....
culmella auct.

1305 **tristella** D. & S. ....

1306 **inquinatella** D. & S. ....

1307 **latistria** Haw. ....

*1308 **poliellus** Treit. ....

1309 **geniculea** Haw. ....

CATOPTRIA Hb.

1310 **permutatella** H.–S. ....

*1311 **osthelderi** Latt. ....

*1312 **speculalis** Hb. ....

1313 **pinella** Linn. ....

1314 **margaritella** D. & S. ....

1315 **furcatellus** Zett. ....

1316 **falsella** D. & S. ....

1317 **verellus** Zinck. ....

*1318 **lythargyrella** Hb. ....

CHRYSOCRAMBUS Blesz.

*1319 **linetella** Fabr. ....

*1320 **craterella** Scop. ....

THISANOTIA Hb.

1321 **chrysonuchella** Scop. ....

PEDIASIA Hb.

1322 **fascelinella** Hb. ....

1323 **contaminella** Hb. ....

1324 **aridella** Thunb. ....

PLATYTES Guen.

1325 **alpinella** Hb. ....

1326 **cerussella** D. & S. ....

ANCYLOLOMIA Hb.

*1327 **tentaculella** Hb. ....

SCHOENOBIINAE

SCHOENOBIUS Dup.

1328 **gigantella** D. & S. ....

1329 **forficella** Thunb. ....

DONACAULA Meyr.

1330 **mucronellus** D. & S. ....

ACENTRIA Steph.

1331 **nivea** Ol. ....
Water Veneer

SCOPARIINAE

SCOPARIA Haw.

1332 **subfusca** Haw. ....
cembrella auct.

1333 **pyralella** D. & S. ....
arundinata Thunb.

1334 **ambigualis** Treit. ....

1334a **basistrigalis** Knaggs

1335 **ulmella** Knaggs ....

EUDONIA Billb.

1336 **pallida** Curt. ....

1337 **alpina** Curt. ....

1338 **crataegella** Hb. ....

1339 **murana** Curt. ....

1340 **truncicolella** Stt. ....

1341 **lineola** Curt. ....

1342 **angustea** Curt. ....

1343 **vandaliella** H.–S. ....
resinella auct.
resinea auct.

1344 **mercurella** Linn. ....

NYMPHULINAE

NYMPHULA Schr.

1345 **nymphaeata** Linn. ....
Brown China-mark

*1346 **enixalis** Swinh. ....

*1347 **melagynalis** Agass. ....

PARAPOYNX Hb.

1348 **stratiotata** Linn. ....
Ringed China-mark

*1349 **obscuralis** Grote ....

1350 **stagnata** Don. ....
Beautiful China-mark

*1351 **diminutalis** Snell. ....

OLIGOSTIGMA Guen.

*1352 **angulipennis** Hamps. ....

*1353 **bilinealis** Snell. ....

*1353a **polydectalis** Walk. ....

CATACLYSTA Hb.

1354 **lemnata** Linn. ....
Small China-mark

SYNCLITA Led.

*1355 **obliteralis** Walk. ....

EVERGESTINAE

EVERGESTIS Hb.

1356 **forficalis** Linn. ....
Garden Pebble

1357 **extimalis** Scop. ....

1358 **pallidata** Hufn. ....

ODONTIINAE

CYNAEDA Hb.

1359 **dentalis** D. & S. ....

GLAPHYRIINAE

HELLULA Guen.

*1360 **undalis** Fabr. ....

PYRAUSTINAE

PYRAUSTA Schr.

1361 **aurata** Scop. ....

1362 **purpuralis** Linn. ....

1363 **ostrinalis** Hb. ....

1364 **sanguinalis** Linn. ....

1365 **cespitalis** D. & S. ....

1366 **nigrata** Scop. ....

1367 **cingulata** Linn. ....

MARGARITIA Steph.

1368 **sticticalis** Linn. ....

URESIPHITA Hb.

*1369 **limbalis** D. & S. ....

SITOCHROA Hb.

1370 **palealis** D. & S. ....

1371 **verticalis** Linn. ....

PARACORSIA Marion

1372 **repandalis** D. & S. ....

MICROSTEGA Meyr.

1373 **pandalis** Hb. ....
Bordered Pearl

1374 **hyalinalis** Hb. ....

OSTRINIA Hb.

1375 **nubilalis** Hb. ....
European Corn-borer

EURRHYPARA Hb.

1376 **hortulata** Linn. ....
Small Magpie

1377 **lancealis** D. & S. ....

1378 **coronata** Hufn. ....

1379 **terrealis** Treit. ....

1380 **perlucidalis** Hb. ....

ANANIA Hb.

1381 **funebris** Ström ....

1382 **verbascalis** D. & S. ....

1383 **pulveralis** Hb. ....

1384 **stachydalis** Germ. ....

EBULEA Doubl.

1385 **crocealis** Hb. ....

OPSIBOTYS Warr.

1386 **fuscalis** D. & S. ....

NASCIA Curt.

1387 **cilialis** Hb. ....

UDEA Guen.

1388 **elutalis** D. & S. ....

1359 **fulvalis** Hb. ....

1390 **prunalis** D. & S. ....

1391 **decrepitalis** H.–S. ....

1392 **olivalis** D. & S.

1393 **uliginosalis** Steph. ....

*1394 **alpinalis** D. & S. ....

1395 **ferrugalis** Hb. ....

MECYNA Doubl.

1396 **flavalis** D. & S.
ssp. **flaviculalis** Carad. ....

1397 **asinalis** Hb. ....

NOMOPHILA Hb.

1398 **noctuella** D. & S. ....
Rush Veneer

DOLICHARTHRIA Steph.

1399 **punctalis** D. & S. ....

ANTIGASTRA Steph.

*1400 **catalaunalis** Dup. ....

MARUCA Walk.

*1401 **testulalis** Geyer ....
Mung Moth

DIASEMIA Hb.

1402 **litterata** Scop. ....

DIASEMIOPSIS Munr.

*1403 **ramburialis** Dup. ....

HYMENIA Hb.

*1404 **recurvalis** Fabr. ....

PLEUROPTYA Meyr.

1405 **ruralis** Scop. ....
Mother of Pearl

HERPETOGRAMMA Led.

*1406 **centrostrigalis** Steph. ....

*1407 **aegrotalis** Zell. ....

PALPITA Hb.

*1408 **unionalis** Hb. ....

DIAPHANIA Hb.

*1409 **lucernalis** Hb. ....

AGROTERA Schr.

1410 **nemoralis** Scop. ....

LEUCINODES Guen.

*1411 **vagans** Tutt ....

DARABA Walk.

*1412 **laisalis** Walk. ....

PYRALINAE

HYPSOPYGIA Hb.

1413 **costalis** Fabr. ....
Gold Triangle

SYNAPHE Hb.

1414 **punctalis** Fabr. ....
angustalis D. & S.

ORTHOPYGIA Rag.

1415 **glaucinalis** Linn. ....

PYRALIS Linn.

1416 **lienigialis** Zell. ....

1417 **farinalis** Linn. ....
Meal Moth

*1418 **manihotalis** Guen. ....

*1419 **pictalis** Curt. ....
Painted Meal Moth

AGLOSSA Latr.

1420 **caprealis** Hb. ....
Small Tabby

1421 **pinguinalis** Linn. ....
Large Tabby

*1422 **dimidiata** Haw. ....
Tea Tabby

*1423 **ocellalis** Led. ....

ENDOTRICHA Zell.

1424 **flammealis** D. & S. ....

GALLERIINAE

GALLERIA Fabr.

1425 **mellonella** Linn. ....
Wax Moth

ACHROIA Hb.

1426 **grisella** Fabr. ....
Lesser Wax Moth

CORCYRA Rag.

1427 **cephalonica** Stt. ....
Rice Moth

APHOMIA Hb.

1428 **sociella** Linn. ....
Bee Moth

MELISSOBLAPTES Zell.

1429 **zelleri** Joann. ....

PARALIPSA Butl.

1430 **gularis** Zell. ....
Stored Nut Moth

ARENIPSES Hamps.

*1431 **sabella** Hamps. ....

PHYCITINAE

ANERASTIA Hb.

1432 **lotella** Hb. ....

CRYPTOBLABES Zell.

1433 **bistriga** Haw. ....

*1434 **gnidiella** Mill. ....

ACROBASIS Zell.

1435 **tumidana** D. & S. ....

1436 **repandana** Fabr. ....
tumidella Zinck.

1437 **consociella** Hb. ....

EURHODOPE Hb.

1438 **suavella** Zinck. ....

1439 **advenella** Zinck. ....

1440 **marmorea** Haw. ....

ONCOCERA Steph.

1441 **semirubella** Scop. ....

1442 **palumbella** D. & S. ....

1443 **genistella** Dup. ....

1444 **obductella** Zell. ....

1445 **formosa** Haw. ....

SALEBRIOPSIS Hann.

1446 **albicilla** H.–S. ....

NEPHOPTERIX Hb.

1447 **hostilis** Steph. ....

SALAGIA Hb.

*1448 **argyrella** D. & S. ....

MICROTHRIX Rag.

1449 **similella** Zinck. ....

METRIOSTOLA Rag.

1450 **betulae** Goeze ....

PYLA Grote

1451 **fusca** Haw. ....

PHYCITA Curt.

1452 **roborella** D. & S. ....

PIMA Hulst

1453 **boisduvaliella** Guen. ....

DIORYCTRIA Zell.

1454 **abietella** D. & S. ....

1455 **mutatella** Fuchs ....

EPISCHNIA Hb.

1456 **bankesiella** Rich. ....

HYPOCHALCIA Hb.

1457 **ahenella** D. & S. ....

MYELOIS Hb.

1458 **cribrella** Hb. ....
Thistle Ermine

1459 **cirrigerella** Zinck. ....

ECTOMYELOIS Heinr.

1460 **ceratoniae** Zell. ....
Locust Bean Moth

ASSARA Walk.

1461 **terebrella** Zinck. ....

PEMPELIA Hb.

1462 **dilutella** Hb. ....

1463 **ornatella** D. & S. ....

GYMNANCYLA Zell.

1464 **canella** D. & S. ....

ALISPA Zell.

1465 **angustella** Hb. ....

MUSSIDIA Rag.

*1466 **nigrivenella** Rag. ....

ANCYLOSIS Zell.

1467 **oblitella** Zell. ....

NYCTEGRETIS Zell.

1468 **achatinella** Hb. ....

EUZOPHERA Zell.

1469 **cinerosella** Zell. ....

1470 **pinguis** Haw. ....

*1471 **osseatella** Treit. ....

*1472 **bigella** Zell. ....

EPHESTIA Guen.

1473 **elutella** Hb. ....
Cacao Moth

1474 **parasitella** Stdgr
ssp. **unicolorella** Stdgr ....

1475 **kuehniella** Zell. ....
Mediterranean Flour Moth

1476 **cautella** Walk. ....
Dried Currant Moth

1477 **figulilella** Gregs. ....
Raisin Moth

1478 **calidella** Guen. ....
Dried Fruit Moth

PLODIA Guen.

1479 **interpunctella** Hb. ....
Indian Meal Moth

HOMOEOSOMA Curt.

1480 **nebulella** D. & S. ....

1481 **sinuella** Fabr. ....

1482 **nimbella** Dup. ....

PHYCITODES Hamps.

1483 **binaevella** Hb. ....

1484 **saxicola** Vaugh. ....

1485 **carlinella** Hein. ....

APOMYELOIS Heinr.

1486 **bistriatella** Hulst
ssp. **neophanes** Durr. ....

PTEROPHORIDAE
AGDISTINAE

AGDISTIS Hb.

1487 **staticis** Mill. ....

1488 **bennetii** Curt. ....

PLATYPTILIINAE

OXYPTILUS Zell.

1489 **pilosellae** Zell. ....

1490 **parvidactylus** Haw. ....

CROMBRUGGHIA Tutt

1491 **distans** Zell. ....

1492 **laetus** Zell. ....

BUCKLERIA Tutt

1493 **paludum** Zell. ....

CAPPERIA Tutt

1494 **britanniodactyla** Gregs. ....

MARASMARCHA Meyr.

1495 **lunaedactyla** Haw. ....

CNAEMIDOPHORUS Wall.

1496 **rhododactyla** D. & S. ....

AMBLYPTILIA Hb.

1497 **acanthadactyla** Hb. ....

1498 **punctidactyla** Haw. ....

PLATYPTILIA Hb.

1499 **tesseradactyla** Linn. ....

1500 **calodactyla** D. & S. ....

1501 **gonodactyla** D. & S. ....

1502 **isodactylus** Zell. ....

1503 **ochrodactyla** D. & S. ....

1504 **pallidactyla** Haw. ....

STENOPTILIA Hb.

1505 **graphodactyla** Treit. ....
pneumonanthes Bütt.

1506 **saxifragae** Fletch. ....

1507 **zophodactylus** Dup. ....

1508 **bipunctidactyla** Scop. ....

1509 **pterodactyla** Linn. ....

PTEROPHORINAE

PTEROPHORUS Schäff.

1510 **tridactyla** Linn. ....

1511 **fuscolimbatus** Dup.
icterodactylus Mann
ssp. **phillipsi** Huggins ....

1512 **baliodactylus** Zell.

1513 **pentadactyla** Linn. ....
White Plume Moth

1514 **galactodactyla** D. & S. ....

1515 **spilodactylus** Curt. ....

PSELNOPHORUS Wall.

1516 **heterodactyla** Müll. ....

ADAINA Tutt

1517 **microdactyla** Hb. ....

LEIOPTILUS Wall.

1518 **lienigianus** Zell. ....

1519 **carphodactyla** Hb. ....

1520 **osteodactylus** Zell. ....

1521 **chrysocomae** Rag. ....
bowesi Whalley

1522 **tephradactyla** Hb. ....

OIDAEMATOPHORUS Wall.

1523 **lithodactyla** Treit. ....

EMMELINA Tutt

1524 **monodactyla** Linn. ....

HESPERIIDAE
HESPERIINAE

CARTEROCEPHALUS Led.

1525 **palaemon** Pallas ....
Chequered Skipper

THYMELICUS Hb.

1526 **sylvestris** Poda ....
Small Skipper

1527 **lineola** Ochs. ....
Essex Skipper

1528 **acteon** Rott. ....
Lulworth Skipper

HESPERIA Fabr.

1529 **comma** Linn. ....
Silver-spotted Skipper

HYLEPHILA Billb.

*1530 **phyleus** Drury ....
Fiery Skipper

OCHLODES Scudd.

1531 **venata** Brem. & Grey
ssp. **faunus** Turati ....
Large Skipper

PYRGINAE

ERYNNIS Schr.

1532 **tages** Linn.
ssp. **tages** Linn. ....
Dingy Skipper

tages Linn.
ssp. **baynesi** Huggins (Ireland)..
Dingy Skipper

CARCHARODUS Hb.

*1533 **alceae** Esp. ....
Mallow Skipper

PYRGUS Hb.

1534 **malvae** Linn. ....
Grizzled Skipper

*1535 **armoricanus** Ob. ....
Oberthür's Grizzled Skipper

PAPILIONIDAE
PARNASSIINAE

PARNASSIUS Lat.

*1536 **apollo** Linn. ....
The Apollo

*1537 **phoebus** Fabr. ....
Small Apollo

ZERYNTHIINAE

PARNALIUS Raf.

*1538 **rumina** Linn. ....
Spanish Festoon

PAPILIONINAE

PAPILIO Linn.

1539 **machaon** Linn.
ssp. **britannicus** Seitz (British) ....
The Swallowtail

machaon Linn.
ssp. **bigeneratus** Ver. (Continental) ....
The Swallowtail

IPHICLIDES Hb.

*1540 **podalirius** Scop. ....
Scarce Swallowtail

PIERIDAE
DISMORPHIINAE

LEPTIDEA Billb.

1541 **sinapis** Linn.
ssp. **sinapis** Linn. ....
Wood White

sinapis Linn.
ssp. **juvernica** Will. (Ireland)..
Wood White

COLIADINAE

COLIAS Fabr.

*1542 **palaeno** Linn. ....
Moorland Clouded Yellow

1543 **hyale** Linn. ....
Pale Clouded Yellow

1544 **australis** Ver. ....
Berger's Clouded Yellow

1545 **croceus** Geoffr. ....
Clouded Yellow

GONEPTERYX Leach

1546 **rhamni** Linn.
ssp. **rhamni** Linn. ....
The Brimstone

rhamni Linn.
ssp. **gravesi** Huggins (Ireland)..
The Brimstone

*1547 **cleopatra** Linn. ....
The Cleopatra

PIERINAE

APORIA Hb.

†1548 **crataegi** Linn. ....
Black-veined White

PIERIS Schr.

1549 **brassicae** Linn. ....
Large White

1550 **rapae** Linn. ....
Small White

1551 **napi** Linn.
ssp. **sabellicae** Steph. ....
Green-veined White

napi Linn.
ssp. **thomsoni** Warr. (Scotland: Perth–Fife–Stirlingshire).....
Green-veined White

napi Linn.
ssp. **britannica** Ver. (Ireland)..
Green-veined White

PONTIA Fabr.

1552 **daplidice** Linn. ....
Bath White

ANTHOCHARIS Boisd.

1553 **cardamines** Linn.
ssp. **britannica** Ver. ....
Orange-tip

cardamines Linn.
ssp. **hibernica** Will. (Ireland)..
Orange-tip

EUCHLOE Hb.

*1554 **ausonia** Hb. ....
Dappled White

LYCAENIDAE
THECLINAE

CALLOPHRYS Billb.

1555 **rubi** Linn. ....
Green Hairstreak

THECLA Fabr.

1556 **betulae** Linn. ....
Brown Hairstreak

QUERCUSIA Ver.

1557 **quercus** Linn. ....
Purple Hairstreak

STRYMONIDIA Tutt

1558 **w-album** Knoch ....
White Letter Hairstreak

1559 **pruni** Linn. ....
Black Hairstreak

RAPALA Moore

*1560 **schistacea** Moore ....
Slate Flash

LYCAENINAE

LYCAENA Fabr.

1561 **phlaeas** Linn.
ssp. **eleus** Fabr. ....
Small Copper

phlaeas Linn.
ssp. **hibernica** Goods. (Ireland)..
Small Copper

†1562 **dispar** Haw.
ssp. **dispar** Haw. ....
Large Copper

* **dispar** Haw.
ssp. **batavus** Ob. (Introduced) ....
Large Copper

* **dispar** Haw.
ssp. **rutilus** Werneb. (Introduced) ....
Large Copper

*1563 **virgaureae** Linn. ....
Scarce Copper

*1564 **tityrus** Poda ....
Sooty Copper

*1565 **alciphron** Rott. ....
Purple-shot Copper

*1566 **hippothoe** Linn. ....
Purple-edged Copper

LAMPIDES Hb.

1567 **boeticus** Linn. ....
Long-tailed Blue

SYNTARUCUS Butl.

*1568 **pirithous** Linn. ....
Lang's Short-tailed Blue

CUPIDO Schr.

1569 **minimus** Fuess. ....
Small Blue

EVERES Hb.

1570 **argiades** Pallas ....
Short-tailed Blue

PLEBEJUS Kluk

1571 **argus** Linn.
ssp. **argus** Linn. ....
Silver-studded Blue

argus Linn.
ssp. **cretaceus** Tutt (Kent)..
Silver-studded Blue

argus Linn.
ssp. **masseyi** Tutt (Lancs.; Westmorland).......................
Silver-studded Blue

argus Linn.
ssp. **caernensis** Thomps. (Caernarvonshire)..............................
Silver-studded Blue

ARICIA R.L.

1572 **agestis** D. & S. ....
Brown Argus

1573 **artaxerxes** Fabr.
ssp. **artaxerxes** Fabr. (Scotland) ....
Northern Brown Argus

artaxerxes Fabr.
ssp. **salmacis** Steph. (N. England) ....
Castle Eden Argus

POLYOMMATUS Lat.

1574 **icarus** Rott.
ssp. **icarus** Rott. ....
Common Blue

icarus Rott.
ssp. **mariscolore** Kane (Ireland)...
Common Blue

LYSANDRA Hemm.

1575 **coridon** Poda ....
Chalk Hill Blue

1576 **bellargus** Rott. ....
Adonis Blue

PLEBICULA Higgins

1577 **dorylas** D. & S. ....
Turquoise Blue

CYANIRIS Dalm.

1578 **semiargus** Rott. ....
Mazarine Blue

GLAUCOPSYCHE Scudd.

1579 **alexis** Poda ....
Green-underside Blue

CELASTRINA Tutt

1580 **argiolus** Linn.
ssp. **britanna** Ver. ....
Holly Blue

MACULINEA Eecke

1581 **arion** Linn.
ssp. **eutyphron** Fruh. ....
Large Blue

NEMEOBIIDAE

HAMEARIS Hb.

1582 **lucina** Linn. ....
Duke of Burgundy Fritillary

NYMPHALIDAE

DRYAS Hb.

*1583 **julia** Fabr.
ssp. **delila** Fabr.
The Julia ....

LADOGA Moore

1584 **camilla** Linn.
White Admiral ....

APATURA Fabr.

1585 **iris** Linn.
Purple Emperor ....

JUNONIA Hb.

*1586 **villida** Fabr.
(**hampstediensis** Jerm.)
Albin's Hampstead Eye ....

*1587 **oenone** Linn.
Blue Pansy ....

COLOBURA Billb.

*1588 **dirce** Linn.
The Zebra ....

HYPANARTIA Hb.

*1589 **lethe** Fabr.
Small Brown Shoemaker ....

VANESSA Fabr.

1590 **atalanta** Linn.
Red Admiral ....

CYNTHIA Fabr.

1591 **cardui** Linn.
Painted Lady ....

1592 **virginiensis** Drury
American Painted Lady ....

AGLAIS Dalm.

1593 **urticae** Linn.
Small Tortoiseshell ....

NYMPHALIS Kluk

1594 **polychloros** Linn.
Large Tortoiseshell ....

*1595 **xanthomelas** D. & S.
Scarce Tortoiseshell ....

1596 **antiopa** Linn.
Camberwell Beauty ....

INACHIS Hb.

1597 **io** Linn.
The Peacock ....

POLYGONIA Hb.

1598 **c-album** Linn.
The Comma ....

ARASCHNIA Hb.

*1599 **levana** Linn.
European Map ....

BOLORIA Moore

1600 **selene** D. & S.
ssp. **selene** D. & S. ....
Small Pearl-bordered Fritillary

selene D. & S.
ssp. **insularum** Harr. (Hebrides)..
Small Pearl-bordered Fritillary

1601 **euphrosyne** Linn.
Pearl-bordered Fritillary ....

*1602 **dia** Linn.
Weaver's Fritillary ....

ARGYNNIS Fabr.

1603 **lathonia** Linn.
Queen of Spain Fritillary ....

*1604 **aphrodite** Fabr.
Aphrodite Fritillary ....

*1605 **niobe** Linn.
Niobe Fritillary ....

1606 **adippe** D. & S.
ssp. **vulgoadippe** Ver. ....
High Brown Fritillary

1607 **aglaja** Linn.
ssp. **aglaja** Linn. ....
Dark Green Fritillary

aglaja Linn.
ssp. **scotica** Watk. (Scotland) ....
Dark Green Fritillary

1608 **paphia** Linn.
Silver-washed Fritillary ....

*1609 **pandora** D. & S.
Mediterranean Fritillary ....

EURODRYAS Higgins

1610 **aurinia** Rott.
ssp. **aurinia** Rott. ....
Marsh Fritillary

aurinia Rott.
ssp. **hibernica** Birch. (Ireland)..
Marsh Fritillary

MELITAEA Fabr.

*1611 **didyma** Esp. ....
Spotted Fritillary

1612 **cinxia** Linn. ....
Glanville Fritillary

MELLICTA Billb.

1613 **athalia** Rott. ....
Heath Fritillary

SATYRIDAE

PARARGE Hb.

1614 **aegeria** Linn.
ssp. **tircis** Butl. ....
Speckled Wood

aegeria Linn.
ssp. **oblita** Harr. (Inner Hebrides: Canna; Rhum)..........
Speckled Wood

aegeria Linn.
ssp. **insula** How. (Scilly Is.)..
Speckled Wood

LASIOMMATA
Humph. & Westw.

1615 **megera** Linn. ....
The Wall

*1616 **maera** Linn. ....
Large Wall

EREBIA Dalm.

1617 **epiphron** Knoch
ssp. **mnemon** Haw. ....
Mountain Ringlet

epiphron Knoch
ssp. **scotica** Cooke (Scotland)..
Mountain Ringlet

1618 **aethiops** Esp. ....
Scotch Argus

*1619 **ligea** Linn. ....
Arran Brown

MELANARGIA Meig.

1620 **galathea** Linn.
ssp. **serena** Ver. ....
Marbled White

HIPPARCHIA Fabr.

1621 **semele** Linn.
ssp. **semele** Linn. ....
The Grayling

semele Linn.
ssp. **thyone** Thomps. (N. Wales)..
The Grayling

semele Linn.
ssp. **atlantica** Harr. (Inner Hebrides)..................................
The Grayling

semele Linn.
ssp. **scota** Ver. (E. Scotland).......................................
The Grayling

semele Linn.
ssp. **hibernica** How. (Ireland)..
The Grayling

semele Linn.
ssp. **clarensis** Lattin (W. Ireland: Co. Clare).........................
The Grayling

*1622 **fagi** Scop. ....
Woodland Grayling

*1623 **briseis** Linn. ....
The Hermit

ARETHUSANA Lesse

*1624 **arethusa** D. & S. ....
False Grayling

PYRONIA Hb.

1625 **tithonus** Linn.
ssp. **britanniae** Ver. ....
The Gatekeeper

MANIOLA Schr.

1626 **jurtina** Linn.
ssp. **insularis** Thompson ....
Meadow Brown

jurtina Linn.
ssp. **iernes** Graves (Ireland)..
Meadow Brown

jurtina Linn.
ssp. **cassiteridum** Graves (Scilly Is.)..
Meadow Brown

jurtina Linn.
ssp. **splendida** White (W. Scotland; Hebrides).......................
Meadow Brown

COENONYMPHA Hb.

1627 **pamphilus** Linn.
ssp. **pamphilus** Linn. ....
Small Heath

pamphilus Linn.
ssp. **rhoumensis** Harr. (Inner Hebrides)..........
Small Heath

1628 **tullia** Müll.
ssp. **davus** Fabr. (N. & C. England (part))..........
Large Heath

tullia Müll.
ssp. **polydama** Haw. (S. Scotland; N. England (part); N. Wales; Ireland)..........
Large Heath

tullia Müll.
ssp. **scotica** Stdgr (Scotland)..........
Large Heath

APHANTOPUS Wall.

1629 **hyperantus** Linn. ....
The Ringlet

DANAIDAE

DANAUS Kluk

1630 **plexippus** Linn. ....
The Milkweed

LASIOCAMPIDAE

POECILOCAMPA Steph.

1631 **populi** Linn. ....
December Moth

TRICHIURA Steph.

1632 **crataegi** Linn. ....
Pale Eggar

ERIOGASTER Germ.

1633 **lanestris** Linn. ....
Small Eggar

MALACOSOMA Hb.

1634 **neustria** Linn. ....
The Lackey

1635 **castrensis** Linn. ....
Ground Lackey

LASIOCAMPA Schr.

1636 **trifolii** D. & S.
ssp. **trifolii** D. & S. ....
Grass Eggar

trifolii D. & S.
ssp. **flava** C.–Hunt (Kent coast) ....
Grass Eggar

1637 **quercus** Linn.
ssp. **quercus** Linn. ....
Oak Eggar

quercus Linn.
ssp. **callunae** Palm. ....
Northern Eggar

MACROTHYLACIA Ramb.

1638 **rubi** Linn. ....
Fox Moth

DENDROLIMUS Germ.

1639 **pini** Linn. ....
Pine-tree Lappet

PHILUDORIA Kirby

1640 **potatoria** Linn. ....
The Drinker

PHYLLODESMA Hb.

1641 **ilicifolia** Linn. ....
Small Lappet

GASTROPACHA Ochs.

1642 **quercifolia** Linn. ....
The Lappet

SATURNIIDAE

SATURNIA Schr.

1643 **pavonia** Linn. ....
Emperor Moth

ENDROMIDAE

ENDROMIS Ochs.

1644 **versicolora** Linn. ....
Kentish Glory

DREPANIDAE

FALCARIA Haw.

1645 **lacertinaria** Linn. ....
Scalloped Hook-tip

DREPANA Schr.

1646 **binaria** Hufn. ....
Oak Hook-tip

1647 **cultraria** Fabr. ....
Barred Hook-tip

1648 **falcataria** Linn.
ssp. **falcataria** Linn. ....
Pebble Hook-tip

falcataria Linn.
ssp. **scotica** Byt.–Salz (Scotland) ....
Pebble Hook-tip

*1649 **curvatula** Borkh. ....
Dusky Hook-tip

SABRE Bode

1650 **harpagula** Esp. ....
Scarce Hook-tip

CILIX Leach

1651 **glaucata** Scop. ....
Chinese Character

THYATIRIDAE

THYATIRA Ochs.

1652 **batis** Linn. ....
Peach Blossom

HABROSYNE Hb.

1653 **pyritoides** Hufn. ....
Buff Arches

TETHEA Ochs.

1654 **ocularis** Linn.
ssp. **octogesimea** Hb. ....
Figure of Eighty

1655 **or** D. & S.
ssp. **or** D. & S. ....
Poplar Lutestring

or D. & S.
ssp. **scotica** Tutt (Scotland) ....
Poplar Lutestring

or D. & S.
ssp. **hibernica** Turn. (Ireland) ....
Poplar Lutestring

TETHEELLA Werny

1656 **fluctuosa** Hb. ....
Satin Lutestring

OCHROPACHA Wall.

1657 **duplaris** Linn. ....
Common Lutestring

CYMATOPHORIMA Spuler

1658 **diluta** D. & S.
ssp. **hartwiegi** Reisser ....
Oak Lutestring

ACHLYA Billb.

1659 **flavicornis** Linn.
ssp. **galbanus** Tutt ....
Yellow Horned

flavicornis Linn.
ssp. **scotica** Tutt (Scotland) ....
Yellow Horned

POLYPLOCA Hb.

1660 **ridens** Fabr. ....
Frosted Green

GEOMETRIDAE
ARCHIEARINAE

ARCHIEARIS Hb.

1661 **parthenias** Linn. ....
Orange Underwing

1662 **notha** Hb. ....
Light Orange Underwing

OENOCHROMINAE

ALSOPHILA Hb.

1663 **aescularia** D. & S. ....
March Moth

GEOMETRINAE

APLASTA Hb.

1664 **ononaria** Fuessl. ....
Rest Harrow

PSEUDOTERPNA Hb.

1665 **pruinata** Hufn.
ssp. **atropunctaria** Walk. ....
Grass Emerald

GEOMETRA Linn.

1666 **papilionaria** Linn. ....
Large Emerald

COMIBAENA Hb.

1667 **bajularia** D. & S. ....
pustulata Hufn.
Blotched Emerald

THETIDIA Boisd.

1668 **smaragdaria** Fabr.
ssp. **maritima** Prout ....
Essex Emerald

HEMITHEA Dup.

1669 **aestivaria** Hb. ....
Common Emerald

CHLORISSA Steph.

1670 **viridata** Linn. ....
Small Grass Emerald

CHLOROCHLAMYS Hulst

*1671 **chloroleucaria** Guen. ....
Blackberry Looper

THALERA Hb.

1672 **fimbrialis** Scop. ....
Sussex Emerald

HEMISTOLA Warr.

1673 **chrysoprasaria** Esp. ....
Small Emerald

JODIS Hb.

1674 **lactearia** Linn. ....
Little Emerald

STERRHINAE

CYCLOPHORA Hb.

1675 **pendularia** Cl. ....
Dingy Mocha

1676 **annulata** Schulze
The Mocha

1677 **albipunctata** Hufn. ....
Birch Mocha

1678 **puppillaria** Hb. ....
Blair's Mocha

1679 **porata** Linn. ....
False Mocha

1680 **punctaria** Linn. ....
Maiden's Blush

1681 **linearia** Hb. ....
Clay Triple-lines

TIMANDRA Dup.

1682 **griseata** Peters. ....
Blood-vein

SCOPULA Schr.

1683 **immorata** Linn. ....
Lewes Wave

1684 **nigropunctata** Hufn. ....
Sub-angled Wave

*1685 **virgulata** D. & S. ....
Streaked Wave

*1686 **decorata** D. & S. ....
Middle Lace Border

1687 **ornata** Scop. ....
Lace Border

1688 **rubiginata** Hufn. ....
Tawny Wave

1689 **marginepunctata** Goeze ....
Mullein Wave

1690 **imitaria** Hb. ....
Small Blood-vein

1691 **emutaria** Hb. ....
Rosy Wave

1692 **immutata** Linn. ....
Lesser Cream Wave

1693 **floslactata** Haw.
ssp. **floslactata** Haw. ....
Cream Wave

floslactata Haw.
ssp. **scotica** Cock. (Scotland) ....
Cream Wave

1694 **ternata** Schr. ....
Smoky Wave

*1695 **limboundata** Haw. ....
Large Lace Border

IDAEA Treit.

1696 **ochrata** Scop.
ssp. **cantiata** Prout ....
Bright Wave

1697 **serpentata** Hufn. ....
Ochraceous Wave

1698 **muricata** Hufn. ....
Purple-bordered Gold

1699 **vulpinaria** H.-S.
ssp. **atrosignaria** Lempke ....
Least Carpet

*1700 **laevigata** Scop. ....
Strange Wave

1701 **sylvestraria** Hb. ....
Dotted Border Wave

1702 **biselata** Hufn.
Small Fan-footed Wave ....

1703 **inquinata** Scop.
Rusty Wave ....

1704 **dilutaria** Hb.
Silky Wave ....

1705 **fuscovenosa** Goeze
Dwarf Cream Wave ....

1706 **humiliata** Hufn.
Isle of Wight Wave ....

1707 **seriata** Schr.
Small Dusty Wave ....

1708 **dimidiata** Hufn.
Single-dotted Wave ....

1709 **subsericeata** Haw.
Satin Wave ....

1710 **contiguaria** Hb.
ssp. **britanniae** Müll.
Weaver's Wave ....

1711 **trigeminata** Haw.
Treble Brown Spot ....

1712 **emarginata** Linn.
Small Scallop ....

1713 **aversata** Linn.
Riband Wave ....

1714 **degeneraria** Hb.
Portland Ribbon Wave ....

1715 **straminata** Borkh.
Plain Wave ....

RHODOMETRA Meyr.

1716 **sacraria** Linn.
The Vestal ....

LARENTIINAE

LYTHRIA Hb.

*1717 **purpuraria** Linn.
Purple-barred Yellow ....

MESOTYPE Hb.

1718 **virgata** Hufn.
Oblique Striped ....

ORTHONAMA Hb.

1719 **vittata** Borkh.
Oblique Carpet ....

1720 **obstipata** Fabr.
The Gem ....

XANTHORHOE Hb.

1721 **biriviata** Borkh. ....
Balsam Carpet

1722 **designata** Hufn, ...
Flame Carpet

1723 **munitata** Hb.
ssp. **munitata** Hb. ....
Red Carpet

munitata Hb.
ssp. **hethlandica** Prout (Shetland Is)..........
Red Carpet

1724 **spadicearia** D. & S. ....
Red Twin-spot Carpet

1725 **ferrugata** Cl. ....
Dark-barred Twin-spot Carpet

1726 **quadrifasiata** Cl. ....
Large Twin-spot Carpet

1727 **montanata** D. & S.
ssp. **montanata** D. & S. ....
Silver-ground Carpet

montanata D. & S.
ssp. **shetlandica** Weir (Shetland Is)..........
Silver-ground Carpet

1728 **fluctuata** Linn.
ssp. **fluctuata** Linn. ....
Garden Carpet

fluctuata Linn.
ssp. **thules** Prout (Aberdeenshire; Shetland Is) ....
Garden Carpet

SCOTOPTERYX Hb.

*1729 **moeniata** Scop. ....
Fortified Carpet

*1730 **peribolata** Hb. ....
Spanish Carpet

1731 **bipunctaria** D. & S.
ssp. **cretata** Prout ....
Chalk Carpet

1732 **chenopodiata** Linn. ....
Shaded Broad-bar

1733 **mucronata** Scop.
ssp. **umbrifera** Heyd. ....
Lead Belle

mucronata Scop.
ssp. **scotica** Cock. (Scotland) ....
Lead Belle

1734 **luridata** Hufn.
ssp. **plumbaria** Fabr. ....
July Belle

CATARHOE Herb.

1735 **rubidata** D. & S. ....
Ruddy Carpet

1736 **cuculata** Hufn. ....
Royal Mantle

EPIRRHOE Hb.

1737 **tristata** Linn. ....
Small Argent & Sable

1738 **alternata** Müll.
ssp. **alternata** Müll. ....
Common Carpet

alternata Müll.
ssp. **obscurata** South (Hebrides)..
Common Carpet

1739 **rivata** Hb. ....
Wood Carpet

1740 **galiata** D. & S. ....
Galium Carpet

COSTACONVEXA Agenjo

†1741 **polygrammata** Borkh. ....
The Many-lined

CAMPTOGRAMMA Steph.

1742 **bilineata** Linn.
ssp. **bilineata** Linn. ....
Yellow Shell

bilineata Linn.
ssp. **atlantica** Stdgr (Hebrides; Shetland Is).........................
Yellow Shell

bilineata Linn.
ssp. **hibernica** Tutt (W. Ireland) ....
Yellow Shell

bilineata Linn.
ssp. **isolata** Kane (Kerry Coast Is) ....
Yellow Shell

ENTEPHRIA Hb.

1743 **flavicinctata** Hb.
ssp. **flavicinctata** Hb. ....
Yellow-ringed Carpet

flavicinctata Hb.
ssp. **ruficinctata** Guen. (Scotland)..
Yellow-ringed Carpet

1744 **caesiata** D. & S. ....
Grey Mountain Carpet

LARENTIA Treit.

1745 **clavaria** Haw. ....
The Mallow

ANTICLEA Steph.

1746 **badiata** D. & S. ....
Shoulder Stripe

1747 **derivata** D. & S. ....
The Streamer

MESOLEUCA Hb.

1748 **albicillata** Hb. ....
Beautiful Carpet

PELURGA Hb.

1749 **comitata** Linn. ....
Dark Spinach

LAMPROPTERYX Steph.

1750 **suffumata** D. & S. ....
Water Carpet

1751 **otregiata** Metc. ....
Devon Carpet

COSMORHOE Hb.

1752 **ocellata** Linn. ....
Purple Bar

COENOTEPHRIA Prout

1753 **salicata** Hb.
ssp. **latentaria** Curt. ....
Striped Twin-spot Carpet

EULITHIS Hb.

1754 **prunata** Linn. ....
The Phoenix

1755 **testata** Linn. ....
The Chevron

1756 **populata** Linn. ....
Northern Spinach

1757 **mellinata** Fabr. ....
The Spinach

1758 **pyraliata** D. & S. ....
Barred Straw

ECLIPTOPERA Warr.

1759 **silaceata** D. & S. ....
Small Phoenix

CHLOROCLYSTA Hb.

1760 **siterata** Hufn. ....
Red-green Carpet

1761 **miata** Linn. ....
Autumn Green Carpet

1762 **citrata** Linn.
ssp. **citrata** Linn. ....
Dark Marbled Carpet

citrata Linn.
ssp. **pythonissata** Mill. (Shetland Is)..........
Dark Marbled Carpet

1763 **concinnata** Steph. ....
Arran Carpet

1764 **truncata** Hufn. ....
Common Marbled Carpet

CIDARIA Treit.

1765 **fulvata** Forst. ....
Barred Yellow

PLEMYRIA Hb.

1766 **rubiginata** D. & S.
ssp. **rubiginata** D. & S. ....
Blue-bordered Carpet

rubiginata D. & S.
ssp. **plumbata** Curt. (Scotland)..........
Blue-bordered Carpet

THERA Steph.

1767 **firmata** Hb. ....
Pine Carpet

1768 **obeliscata** Hb. ....
Grey Pine Carpet

*1769 **variata** D. & S.
ssp. **variata** D. & S. ....
Spruce Carpet

variata D. & S.
ssp. **britannica** Turn. ....
Spruce Carpet

1770 **cognata** Thunb. ....
Chestnut-coloured Carpet

1771 **juniperata** Linn.
ssp. **juniperata** Linn. ....
Juniper Carpet

juniperata Linn.
ssp. **scotica** White (Scotland)..........
Juniper Carpet

juniperata Linn.
ssp. **orcadensis** Cock. (Orkney Is)..........
Juniper Carpet

EUSTROMA Hb.

1772 **reticulatum** D. & S. ....
Netted Carpet

ELECTROPHAES Prout

1773 **corylata** Thunb. ....
Broken-barred Carpet

COLOSTYGIA Hb.

1774 **olivata** D. & S. ....
Beech-green Carpet

1775 **multistrigaria** Haw. ....
Mottled Grey

1776 **pectinataria** Knoch ....
Green Carpet

HYDRIOMENA Hb.

1777 **furcata** Thunb. ....
July Highflyer

1778 **impluviata** D. & S. ....
May Highflyer

1779 **ruberata** Freyer ....
Ruddy Highflyer

COENOCALPE Hb.

1780 **lapidata** Hb. ....
Slender-striped Rufous

HORISME Hb.

1781 **vitalbata** D. & S. ....
Small Waved Umber

1782 **tersata** D. & S. ....
The Fern

*1783 **aquata** Hb. ....
Cumbrian Umber

MELANTHIA Dup.

1784 **procellata** D. & S. ....
Pretty Chalk Carpet

PAREULYPE Herb.

1785 **berberata** D. & S. ....
Barberry Carpet

SPARGANIA Guen.

1786 **luctuata** D. & S. ....
White-banded Carpet

RHEUMAPTERA Hb.

1787 **hastata** Linn.
ssp. **hastata** Linn. ....
Argent & Sable

hastata Linn.
ssp. **nigrescens** Prout (Scotland) ....
Argent & Sable

1788 **cervinalis** Scop. ....
Scarce Tissue

1789 **undulata** Linn. ....
Scallop Shell

TRIPHOSA Steph.

1790 **dubitata** Linn. ....
The Tissue

PHILEREME Hb.

1791 **vetulata** D. & S. ....
Brown Scallop

1792 **transversata** Hufn.
ssp. **britannica** Lempke ....
Dark Umber

EUPHYIA Hb.

1793 **biangulata** Haw. ....
Cloaked Carpet

1794 **unangulata** Haw. ....
Sharp-angled Carpet

EPIRRITA Hb.

1795 **dilutata** D. & S. ....
November Moth

1796 **christyi** Allen
Pale November Moth

1797 **autumnata** Borkh. ....
Autumnal Moth

1798 **filigrammaria** H.–S. ....
Small Autumnal Moth

OPEROPHTERA Hb.

1799 **brumata** Linn. ....
Winter Moth

1800 **fagata** Scharf. ....
Northern Winter Moth

PERIZOMA Hb.

1801 **taeniatum** Steph. ....
Barred Carpet

1802 **affinitatum** Steph. ....
The Rivulet

1803 **alchemillata** Linn. ....
Small Rivulet

1804 **bifaciata** Haw. ....
Barred Rivulet

1805 **minorata** Treit.
ssp. **ericetata** Steph. ....
Heath Rivulet

1806 **blandiata** D. & S.
ssp. **blandiata** D. & S. ....
Pretty Pinion

blandiata D. & S.
ssp. **perfasciata** Prout (Hebrides)..........
Pretty Pinion

1807 **albulata** D. & S.
ssp. **albulata** D. & S. ....
Grass Rivulet

albulata D. & S.
ssp. **subfasciaria** Bohem. (Shetland Is)..........
Grass Rivulet

1808 **flavofasciata** Thunb. ....
Sandy Carpet

1809 **didymata** Linn.
ssp. **didymata** Linn. ....
Twin-spot Carpet

didymata Linn.
ssp. **hethlandica** Rebel (Shetland Is)..........
Twin-spot Carpet

1810 **sagittata** Fabr. ....
Marsh Carpet

EUPITHECIA Curt.

1811 **tenuiata** Hb. ....
Slender Pug

1812 **inturbata** Hb. ....
Maple Pug

1813 **haworthiata** Doubl. ....
Haworth's Pug

1814 **plumbeolata** Haw. ....
Lead-coloured Pug

1815 **abietaria** Goeze ....
Cloaked Pug

1816 **linariata** D. & S. ....
Toadflax Pug

1817 **pulchellata** Steph.
ssp. **pulchellata** Steph. ....
Foxglove Pug

pulchellata Steph.
ssp. **hebudium** Sheld. (Hebrides) ....
Foxglove Pug

1818 **irriguata** Hb. ....
Marbled Pug

1819 **exiguata** Hb.
ssp. **exiguata** Hb. ....
Mottled Pug

exiguata Hb.
ssp. **muricolor** Prout (E. Aberdeenshire) ....
Mottled Pug

1820 **insigniata** Hb. ....
Pinion-spotted Pug

1821 **valerianata** Hb. ....
Valerian Pug

1822 **pygmaeata** Hb. ....
Marsh Pug

1823 **venosata** Fabr.
ssp. **venosata** Fabr. ....
Netted Pug

venosata Fabr.
ssp. **hebridensis** W. P. Curt. (Hebrides) ....
Netted Pug

venosata Fabr.
ssp. **fumosae** Gregs. (Shetland Is) ....
Netted Pug

venosata Fabr.
ssp. **ochracae** Gregs. (Orkney Is) ....
Netted Pug

venosata Fabr.
ssp. **plumbea** Huggins (Blasket Is) ....
Netted Pug

1824 **egenaria** H.–S. ....
Pauper Pug

1825 **centaureata** D. & S. ....
Lime-speck Pug

1826 **trisignaria** H.–S. ....
Triple-spotted Pug

1827 **intricata** Zett.
ssp. **arceuthata** Freyer ....
Freyer's Pug

intricata Zett.
ssp. **millieraria** Wnuk. ....
Edinburgh Pug

intricata Zett.
ssp. **hibernica** Mere (W. Ireland) ....
Mere's Pug

1828 **satyrata** Hb.
ssp. **satyrata** Hb. ....
Satyr Pug

satyrata Hb.
ssp. **callunaria** Doubl. (Northern moorland) ....
Satyr Pug

satyrata Hb.
ssp. **curzoni** Gregs. (Shetland Is) ....
Satyr Pug

*1829 **cauchiata** Dup. ....

1830 **absinthiata** Cl. ....
Wormwood Pug

1831 **goossensiata** Mab. ....
Ling Pug

1832 **assimilata** Doubl. ....
Currant Pug

1833 **expallidata** Doubl. ....
Bleached Pug

1834 **vulgata** Haw.
ssp. **vulgata** Haw. ....
Common Pug

vulgata Haw.
ssp. **scotica** Cock. (Scotland) ....
Common Pug

vulgata Haw.
ssp. **clarensis** Huggins (W. Ireland) ....
Common Pug

1835 **tripunctaria** H.–S. ....
White-spotted Pug

1836 **denotata** Hb.
ssp. **denotata** Hb. ....
Campanula Pug

denotata Hb.
ssp. **jasioneata** Crewe ....
Jasione Pug

1837 **subfuscata** Haw. ....
Grey Pug

1838 **icterata** Vill.
ssp. **subfulvata** Haw. ....
Tawny Speckled Pug

icterata Vill.
ssp. **cognata** Steph. (Scotland) ....
Tawny Speckled Pug

1839 **succenturiata** Linn. ....
Bordered Pug

1840 **subumbrata** D. & S. ....
Shaded Pug

1841 **millefoliata** Rössl. ....
Yarrow Pug

1842 **simpliciata** Haw. ....
Plain Pug

1843 **distinctaria** H.–S.
ssp. **constrictata** Guen. ....
Thyme Pug

1844 **indigata** Hb. ....
Ochreous Pug

1845 **pimpinellata** Hb. ....
Pimpinel Pug

1846 **nanata** Hb.
ssp. **angusta** Prout ....
Narrow-winged Pug

1847 **extensaria** Freyer
ssp. **occidua** Prout ....
Scarce Pug

*1848 **innotata** Hufn. ....
Angle-barred Pug

1849 **fraxinata** Crewe ....
Ash Pug

*1850 **tamarisciata** Freyer ....
Tamarisk Pug

1851 **virgaureata** Doubl. ....
Golden-rod Pug

1852 **abbreviata** Steph. ....
Brindled Pug

1853 **dodoneata** Guen. ....
Oak-tree Pug

1854 **pusillata** D. & S.
ssp. **anglicata** H.–S. (Kent; Isle of Wight)..........
Juniper Pug

pusillata D. & S.
ssp. **pusillata** D. & S. ....
Juniper Pug

1855 **phoeniceata** Ramb. ....
Cypress Pug

1856 **lariciata** Freyer ....
Larch Pug

1857 **tantillaria** Boisd. ....
Dwarf Pug

CHLOROCLYSTIS Hb.

1858 **v-ata** Haw. ....
The V-Pug

1859 **chloerata** Mab. ....
Sloe Pug

1860 **rectangulata** Linn. ....
Green Pug

1861 **debiliata** Hb. ....
Bilberry Pug

GYMNOSCELIS Mab.

1862 **rufifasciata** Haw. ....
Double-striped Pug

ANTICOLLIX Prout

1863 **sparsata** Treit. ....
Dentated Pug

CHESIAS Treit.

1864 **legatella** D. & S. ....
The Streak

1865 **rufata** Fabr.
ssp. **rufata** Fabr. ....
Broom-tip

rufata Fabr.
ssp. **scotica** Rich. (Scotland) ....
Broom-tip

CARSIA Hb.

1866 **sororiata** Hb.
ssp. **anglica** Prout ....
Manchester Treble-bar

APLOCERA Hb.

1867 **plagiata** Linn.
ssp. **plagiata** Linn. ....
Treble-bar

plagiata Linn.
ssp. **scotica** Rich. (Scotland) ....
Treble-bar

1868 **efformata** Guen. ....
Lesser Treble-bar

1869 **praeformata** Hb. ....
Purple Treble-bar

ODEZIA Boisd.

1870 **atrata** Linn. ....
Chimney Sweeper

LITHOSTEGE Hb.

1871 **griseata** D. & S. ....
Grey Carpet

DISCOLOXIA Warr.

1872 **blomeri** Curt. ....
Blomer's Rivulet

VENUSIA Curt.

1873 **cambrica** Curt.
Welsh Wave ....

EUCHOECA Hb.

1874 **nebulata** Scop.
Dingy Shell ....

ASTHENA Hb.

1875 **albulata** Hufn.
Small White Wave ....

HYDRELIA Hb.

1876 **flammeolaria** Hufn.
Small Yellow Wave ....

1877 **sylvata** D. & S.
Waved Carpet ....

MINOA Treit.

1878 **murinata** Scop.
Drab Looper ....

LOBOPHORA Curt.

1879 **halterata** Hufn.
The Seraphim ....

TRICHOPTERYX Hb.

1880 **polycommata** D. & S.
Barred Tooth-striped ....

1881 **carpinata** Borkh.
Early Tooth-striped ....

PTERAPHERAPTERYX Curt.

1882 **sexalata** Retz.
Small Seraphim ....

ACASIS Dup.

1883 **viretata** Hb.
Yellow-barred Brindle ....

ENNOMINAE

ABRAXAS Leach

1884 **grossulariata** Linn.
The Magpie ....

1885 **sylvata** Scop.
Clouded Magpie ....

*1886 **pantaria** Linn.
Light Magpie ....

LOMASPILIS Hb.

1887 **marginata** Linn. ....
Clouded Border

LIGDIA Guen.

1888 **adustata** D. & S. ....
Scorched Carpet

SEMIOTHISA Hb.

1889 **notata** Linn. ....
Peacock Moth

1890 **alternaria** Hb. ....
Sharp-angled Peacock

*1891 **signaria** Hb. ....

*1892 **praeatomata** Haw. ....
Dingy Angle

1893 **liturata** Cl. ....
Tawny-barred Angle

1894 **clathrata** Linn.
ssp. **clathrata** Linn. ....
Latticed Heath

clathrata Linn.
ssp. **hugginsi** Baynes (Ireland)..
Latticed Heath

1895 **carbonaria** Cl. ....
Netted Mountain Moth

1896 **brunneata** Thunb. ....
Rannoch Looper

1897 **wauaria** Linn. ....
The V-Moth

HYPAGYRTIS Hb.

*1898 **unipunctata** Haw. ....
White Spot

ISTURGIA Hb.

1899 **limbaria** Fabr. ....
Frosted Yellow

NEMATOCAMPA Guen.

*1900 **limbata** Haw. ....
Bordered Chequer

CEPPHIS Hb.

1901 **advenaria** Hb. ....
Little Thorn

PETROPHORA Hb.

1902 **chlorosata** Scop. ....
Brown Silver-line

PLAGODIS Hb.

1903 **pulveraria** Linn. ....
Barred Umber

1904 **dolabraria** Linn. ....
Scorched Wing

PACHYCNEMIA Steph.

1905 **hippocastanaria** Hb. ....
Horse Chestnut

OPISTHOGRAPTIS Hb.

1906 **luteolata** Linn. ....
Brimstone Moth

EPIONE Dup.

1907 **repandaria** Hufn. ....
Bordered Beauty

1908 **paralellaria** D. & S. ....
Dark Bordered Beauty

PSEUDOPANTHERA Hb.

1909 **macularia** Linn. ....
Speckled Yellow

APEIRA Gistl

1910 **syringaria** Linn. ....
Lilac Beauty

ENNOMOS Treit.

1911 **autumnaria** Werneb. ....
Large Thorn

1912 **quercinaria** Hufn. ....
August Thorn

1913 **alniaria** Linn. ....
Canary-shouldered Thorn

1914 **fuscantaria** Haw. ....
Dusky Thorn

1915 **erosaria** D. & S. ....
September Thorn

*1916 **quercaria** Hb. ....
Clouded August Thorn

SELENIA Hb.

1917 **dentaria** Fabr. ....
Early Thorn

1918 **lunularia** Hb. ....
Lunar Thorn

1919 **tetralunaria** Hufn. ....
Purple Thorn

ODONTOPERA Steph.

1920 **bidentata** Cl. ....
Scalloped Hazel

CROCALLIS Treit.

1921 **elinguaria** Linn. ....
Scalloped Oak

OURAPTERYX Leach

1922 **sambucaria** Linn. ....
Swallow-tailed Moth

COLOTOIS Hb.

1923 **pennaria** Hb. ....
Feathered Thorn

ANGERONA Dup.

1924 **prunaria** Linn. ....
Orange Moth

APOCHEIMA Hb.

1925 **hispidaria** D. & S. ....
Small Brindled Beauty

1926 **pilosaria** D. & S. ....
Pale Brindled Beauty

LYCIA Hb.

1927 **hirtaria** Cl. ....
Brindled Beauty

1928 **zonaria** D. & S.
ssp. **britannica** Harr. ....
Belted Beauty

zonaria D. & S.
ssp. **atlantica** Harr. (Hebrides)..
Belted Beauty

1929 **lapponaria** Boisd.
ssp. **scotica** Harr. ....
Rannoch Brindled Beauty

BISTON Leach

1930 **strataria** Hufn. ....
Oak Beauty

1931 **betularia** Linn. ....
Peppered Moth

betularia Linn.
f. **carbonaria** Jord. (melanic form) ....
Peppered Moth

AGRIOPIS Hb.

1932 **leucophaearia** D. & S. ....
Spring Usher

1933 **aurantiaria** Hb. ....
Scarce Umber

1934 **marginaria** Fabr. ....
Dotted Border

ERANNIS Hb.

1935 **defoliaria** Cl. ....
Mottled Umber

MENOPHRA Moore

1936 **abruptaria** Thunb. ....
Waved Umber

PERIBATODES Wehrli

1937 **rhomboidaria** D. & S. ....
Willow Beauty

SELIDOSEMA Hb.

1938 **brunnearia** Vill.
ssp. **scandinaviaria** Stdgr ....
Bordered Grey

brunnearia Vill.
ssp. **tyronensis** Cock. (Ireland: Co. Tyrone) ....
Bordered Grey

CLEORA Curt.

1939 **cinctaria** D. & S.
ssp. **cinctaria** D. & S. ....
Ringed Carpet

cinctaria D. & S.
ssp. **bowesi** Rich. (Scotland) ....
Ringed Carpet

DEILEPTENIA Hb.

1940 **ribeata** Cl. ....
Satin Beauty

ALCIS Curt.

1941 **repandata** Linn.
ssp. **repandata** Linn. ....
Mottled Beauty

repandata Linn.
ssp. **muraria** Curt. (Scotland) ....
Mottled Beauty

repandata Linn.
ssp. **sodorensium** Weir (Hebrides) ....
Mottled Beauty

1942 **jubata** Thunb. ....
Dotted Carpet

BOARMIA Treit.

1943 **roboraria** D. & S. ....
Great Oak Beauty

SERRACA Moore

1944 **punctinalis** Scop. ....
Pale Oak Beauty

CLEORODES Warr.

1945 **lichenaria** Hufn. ....
Brussels Lace

FAGIVORINA Wehrli

†1946 **arenaria** Hufn. ....
Speckled Beauty

ECTROPIS Hb.

1947 **bistortata** Goeze ....
The Engrailed

1948 **crepuscularia** D. & S. ....
Small Engrailed

1949 **consonaria** Hb. ....
Square Spot

1950 **extersaria** Hb. ....
Brindled White-spot

AETHALURA McDunn.

1951 **punctulata** D. & S. ....
Grey Birch

EMATURGA Led.

1952 **atomaria** Linn.
ssp. **atomaria** Linn. ....
Common Heath

atomaria Linn.
ssp. **minuta** Heyd. (Northern heathlands) ....
Common Heath

TEPHRONIA Hb.

*1953 **cremiaria** Freyer ....
Dusky Carpet

BUPALUS Leach

1954 **piniaria** Linn. ....
Bordered White

CABERA Treit.

1955 **pusaria** Linn. ....
Common White Wave

1956 **exanthemata** Scop. ....
Common Wave

LOMOGRAPHA Hb.

1957 **bimaculata** Fabr. ....
White-pinion Spotted

1958 **temerata** D. & S. ....
Clouded Silver

ALEUCIS Guen.

1959 **distinctata** H.–S. ....
Sloe Carpet

THERIA Hb.

1960 **primaria** Haw. ....
rupicapraria auct.
Early Moth

CAMPAEA Lamarck

1961 **margaritata** Linn. ....
Light Emerald

HYLAEA Hb.

1962 **fasciaria** Linn. ....
Barred Red

GNOPHOS Treit.

1963 **obfuscatus** D. & S. ....
Scotch Annulet

1964 **obscuratus** D. & S. ....
The Annulet

PSODOS Treit.

1965 **coracina** Esp. ....
Black Mountain Moth

SIONA Dup.

1966 **lineata** Scop. ....
Black-veined Moth

ASPITATES Treit.

1967 **gilvaria** D. & S.
ssp. **gilvaria** D. & S. ....
Straw Belle

gilvaria D. & S.
ssp. **burrenensis** Cock. (Ireland: Co. Clare) ....
Straw Belle

1968 **ochrearia** Rossi ....
Yellow Belle

DYSCIA Hb.

1969 **fagaria** Thunb. ....
Grey Scalloped Bar

PERCONIA Hb.

1970 **strigillaria** Hb. ....
Grass Wave

SPHINGIDAE
SPHINGINAE

AGRIUS Hb.

1971 **cingulata** Fabr. ....
Pink-spotted Hawk-moth

1972 **convolvuli** Linn. ....
Convolvulus Hawk-moth

ACHERONTIA Lasp.

1973 **atropos** Linn. ....
Death's-head Hawk-moth

MANDUCA Hb.

1974 **quinquemaculatus** Haw. ....
Five-spotted Hawk-moth

1975 **sexta** Linn. ....
Tomato Sphinx

SPHINX Linn.

1976 **ligustri** Linn. ....
Privet Hawk-moth

1977 **drupiferarum** Smith ....
Wild Cherry Sphinx

HYLOICUS Hb.

1978 **pinastri** Linn. ....
Pine Hawk-moth

MIMAS Hb.

1979 **tiliae** Linn. ....
Lime Hawk-moth

SMERINTHUS Latr.

1980 **ocellata** Linn. ....
Eyed Hawk-moth

LAOTHOE Fabr.

1981 **populi** Linn. ....
Poplar Hawk-moth

MACROGLOSSINAE

HEMARIS Dalm.

1982 **tityus** Linn. ....
Narrow-bordered Bee Hawk-moth

1983 **fuciformis** Linn. ....
Broad-bordered Bee Hawk-moth

MACROGLOSSUM Scop.

1984 **stellatarum** Linn. ....
Humming-bird Hawk-moth

DAPHNIS Hb.

1985 **nerii** Linn. ....
Oleander Hawk-moth

HYLES Hb.

1986 **euphorbiae** Linn. ....
Spurge Hawk-moth

1987 **gallii** Rott. ....
Bedstraw Hawk-moth

*1988 **nicaea** Prunner ....
Mediterranean Hawk-moth

*1989 **hippophaes** Esp. ....
Seathorn Hawk-moth

1990 **lineata** Fabr.
ssp. **livornica** Esp. ....
Striped Hawk-moth

DEILEPHILA Lasp.

1991 **elpenor** Linn. ....
Elephant Hawk-moth

1992 **porcellus** Linn. ....
Small Elephant Hawk-moth

HIPPOTION Hb.

1993 **celerio** Linn. ....
Silver-striped Hawk-moth

NOTODONTIDAE

PHALERA Hb.

1994 **bucephala** Linn. ....
Buff-tip

CERURA Schr.

1995 **vinula** Linn. ....
Puss Moth

FURCULA Lam.

1996 **bicuspis** Borkh. ....
Alder Kitten

1997 **furcula** Cl. ....
Sallow Kitten

1998 **bifida** Brahm ....
Poplar Kitten

STAUROPUS Germ.

1999 **fagi** Linn. ....
Lobster Moth

NOTODONTA Ochs.

2000 **dromedarius** Linn. ....
Iron Prominent

*2001 **torva** Hb. ....
Large Dark Prominent

TRITOPHIA Kiriak.

2002 **tritophus** D. & S. ....
Three-humped Prominent

ELIGMODONTA Kiriak.

2003 **ziczac** Linn. ....
Pebble Prominent

HARPYIA Ochs.

*2004 **milhauseri** Fabr. ....

PERIDEA Steph.

2005 **anceps** Goeze ....
Great Prominent

PHEOSIA Hb.

2006 **gnoma** Fabr. ....
Lesser Swallow Prominent

2007 **tremula** Cl. ....
Swallow Prominent

PTILODON Hb.

2008 **capucina** Linn. ....
Coxcomb Prominent

PTILODONTELLA Kiriak.

2009 **cucullina** D. & S. ....
Maple Prominent

ODONTOSIA Hb.

2010 **carmelita** Esp. ....
Scarce Prominent

PTEROSTOMA Germ.

2011 **palpina** Cl. ....
Pale Prominent

LEUCODONTA Stdgr

2012 **bicoloria** D. & S. ....
White Prominent

PTILOPHORA Steph.

2013 **plumigera** D. & S. ....
Plumed Prominent

DRYMONIA Hb.

2014 **dodonaea** D. & S. ....
Marbled Brown

2015 **ruficornis** Hufn. ....
Lunar Marbled Brown

GLUPHISIA Boisd.

2016 **crenata** Esp.
ssp. **vertunea** Bray ....
Dusky Marbled Brown

CLOSTERA Sam.

2017 **pigra** Hufn. ....
Small Chocolate-tip

2018 **anachoreta** D. & S. ....
Scarce Chocolate-tip

2019 **curtula** Linn. ....
Chocolate-tip

DILOBA Boisd.

2020 **caeruleocephala** Linn. ....
Figure of Eight

THAUMETOPOEIDAE

THAUMETOPOEA Hb.

*2021 **pityocampa** D. & S. ....
Pine Processionary

*2022 **processionea** Linn. ....
Oak Processionary

TRICHOCERCUS Steph.

*2023 **sparshalli** Curt. ....
Local Long-tailed Satin

LYMANTRIIDAE

LAELIA Steph.

†2024 **coenosa** Hb. ....
Reed Tussock

ORGYIA Ochs.

2025 **recens** Hb. ....
Scarce Vapourer

2026 **antiqua** Linn. ....
The Vapourer

DASYCHIRA Hb.

2027 **fascelina** Linn. ....
Dark Tussock

2028 **pudibunda** Linn. ....
Pale Tussock

EUPROCTIS Hb.

2029 **chrysorrhoea** Linn. ....
Brown-tail

2030 **similis** Fuess. ....
Yellow-tail

LEUCOMA Hb.

2031 **salicis** Linn. ....
White Satin Moth

ARCTORNIS Germ.

2032 **l-nigrum** Müll. ....
Black V Moth

LYMANTRIA Hb.

2033 **monacha** Linn. ....
Black Arches

2034 **dispar** Linn. ....
Gypsy Moth

ARCTIIDAE
LITHOSIINAE

THUMATHA Walk.

2035 **senex** Hb. ....
Round-winged Muslin

SETINA Schr.

2036 **irrorella** Linn. ....
Dew Moth

MILTOCHRISTA Hb.

2037 **miniata** Forst. ....
Rosy Footman

NUDARIA Haw.

2038 **mundana** Linn. ....
Muslin Footman

ATOLMIS Hb.

2039 **rubricollis** Linn. ....
Red-necked Footman

CYBOSIA Hb.

2040 **mesomella** Linn. ....
Four-dotted Footman

PELOSIA Hb.

2041 **muscerda** Hufn. ....
Dotted Footman

2042 **obtusa** H.–S. ....
Small Dotted Footman

EILEMA Hb.

2043 **sororcula** Hufn. ....
Orange Footman

2044 **griseola** Hb. ....
Dingy Footman

2045 **caniola** Hb. ....
Hoary Footman

2046 **pygmaeola** Doubl.
ssp. **pygmaeola** Doubl. ....
Pigmy Footman

pygmaeola Doubl.
ssp. **pallifrons** Zell. ....
Dungeness Pigmy Footman

2047 **complana** Linn. ....
Scarce Footman

2048 **sericea** Gregs. ....
Northern Footman

2049 **deplana** Esp. ....
Buff Footman

2050 **lurideola** Zinck. ....
Common Footman

LITHOSIA Fabr.

2051 **quadra** Linn. ....
Four-spotted Footman

ARCTIINAE

SPIRIS Hb.

2052 **striata** Linn. ....
Feathered Footman

COSCINIA Hb.

2053 **cribraria** Linn.
ssp. **bivittata** South ....
Speckled Footman

cribraria Linn.
ssp. **arenaria** Lempke (Kent: Dungeness; Sandwich) ....
Speckled Footman

UTETHEISA Hb.

2054 **pulchella** Linn. ....
Crimson Speckled

*2055 **bella** Linn. ....
Beautiful Utetheisa

PARASEMIA Hb.

2056 **plantaginis** Linn.
ssp. **plantaginis** Linn. ....
Wood Tiger

plantaginis Linn.
ssp. **insularum** Seitz (Orkney Is)..
Wood Tiger

ARCTIA Schr.

2057 **caja** Linn. ....
Garden Tiger

2058 **villica** Linn.
ssp. **britannica** Ob. ....
Cream-spot Tiger

DIACRISIA Hb.

2059 **sannio** Linn. ....
Clouded Buff

SPILOSOMA Curt.

2060 **lubricipeda** Linn. ....
White Ermine

2061 **luteum** Hufn. ....
Buff Ermine

2062 **urticae** Esp. ....
Water Ermine

DIAPHORA Steph.

2063 **mendica** Cl. ....
Muslin Moth

PHRAGMATOBIA Steph.

2064 **fuliginosa** Linn.
ssp. **fuliginosa** Linn. ....
Ruby Tiger

fuliginosa Linn.
ssp. **borealis** Stdgr ....
Northern Ruby Tiger

PYRRHARCTIA Pack.

*2065 **isabella** Smith ....
Isabelline Tiger

HALYSIDOTA Hb.

*2066 **moeschleri** Roths. ....

EUPLAGIA Hb.

2067 **quadripunctaria** Poda ....
Jersey Tiger

CALLIMORPHA Latr.

2068 **dominula** Linn. ....
Scarlet Tiger

TYRIA Hb.

2069 **jacobaeae** Linn. ....
The Cinnabar

CTENUCHIDAE
SYNTOMINAE

SYNTOMIS Ochs.

*2070 **phegea** Linn. ....
The Nine-spotted

DYSAUXES Hb.

*2071 **ancilla** Linn. ....
The Handmaid

EUCHROMIINAE

EUCHROMIA Hb.

*2072 **lethe** Fabr. ....
The Basker

ANTICHLORIS Hb.

*2073 **viridis** Druce ....
musicola Cockll

*2074 **caca** Hb. ....
The Docker

NOLIDAE

MEGANOLA Dyar

2075 **strigula** D. & S. ....
Small Black Arches

2076 **albula** D. & S. ....
Kent Black Arches

NOLA Leach

2077 **cucullatella** Linn. ....
Short-cloaked Moth

2078 **confusalis** H.–S. ....
Least Black Arches

2079 **aerugula** Hb. ....
Scarce Black Arches

NOCTUIDAE
NOCTUINAE

EUXOA Hb.

2080 **obelisca** D. & S.
ssp. **grisea** Tutt ....
Square-spot Dart

2081 **tritici** Linn. ....
White-line Dart

2082 **nigricans** Linn. ....
Garden Dart

2083 **cursoria** Hufn. ....
Coast Dart

AGROTIS Ochs.

2084 **cinerea** D. & S. ....
Light Feathered Rustic

2085 **vestigialis** Hufn. ....
Archer's Dart

*2086 **spinifera** Hb. ....
Gregson's Dart

2087 **segetum** D. & S. ....
Turnip Moth

2088 **clavis** Hufn. ....
Heart & Club

2089 **exclamationis** Linn. ....
Heart & Dart

2090 **trux** Hb.
ssp. **lunigera** Steph. ....
Crescent Dart

2091 **ipsilon** Hufn. ....
Dark Sword-grass

2092 **puta** Hb.
ssp. **puta** Hb. ....
Shuttle-shaped Dart

puta Hb.
ssp. **insula** Rich. (Scilly Is) ....
Shuttle-shaped Dart

2093 **ripae** Hb. ....
Sand Dart

*2094 **crassa** Hb. ....
Great Dart

FELTIA Walk.

*2095 **subgothica** Haw. ....
Gothic Dart

*2096 **subterranea** Fabr. ....
Tawny Shoulder

ACTINOTIA Hb.

2097 **polyodon** Cl. ....
Purple Cloud

AXYLIA Hb.

2098 **putris** Linn. ....
The Flame

OCHROPLEURA Hb.

2099 **praecox** Linn. ....
Portland Moth

*2100 **fennica** Tausch. ....
Eversmann's Rustic

*2101 **flammatra** D. & S. ....
Black Collar

2102 **plecta** Linn. ....
Flame Shoulder

EUGNORISMA Bours.

2103 **depuncta** Linn. ....
Plain Clay

STANDFUSSIANA Bours.

2104 **lucernea** Linn. ....
Northern Rustic

RHYACIA Hb.

2105 **simulans** Hufn. ....
Dotted Rustic

*2106 **lucipeta** D. & S. ....

NOCTUA Linn.

2107 **pronuba** Linn. ....
Large Yellow Underwing

2108 **orbona** Hufn. ....
Lunar Yellow Underwing

2109 **comes** Hb. ....
Lesser Yellow Underwing

2110 **fimbriata** Schreb. ....
Broad-bordered Yellow Underwing

2111 **janthina** D. & S. ....
Lesser Broad-bordered Yellow Underwing

2112 **interjecta** Hb.
ssp. **caliginosa** Schaw. ....
Least Yellow Underwing

SPAELOTIS Boisd.

2113 **ravida** D. & S. ....
Stout Dart

GRAPHIPHORA Ochs.

2114 **augur** Fabr. ....
Double Dart

EUGRAPHE Hb.

2115 **subrosea** Steph. ....
Rosy Marsh Moth

PARADIARSIA McDunn.

2116 **sobrina** Dup. ....
Cousin German

2117 **glareosa** Esp.
ssp. **glareosa** Esp. ....
Autumnal Rustic

glareosa Esp.
ssp. **edda** Stdgr (Shetland Is)..
Autumnal Rustic

LYCOPHOTIA Hb.

2118 **porphyrea** D. & S. ....
True Lover's Knot

PERIDROMA Hb.

2119 **saucia** Hb. ....
Pearly Underwing

DIARSIA Hb.

2120 **mendica** Fabr.
ssp. **mendica** Fabr. ....
Ingrailed Clay

mendica Fabr.
ssp. **thulei** Stdgr (Shetland Is)..
Ingrailed Clay

mendica Fabr.
ssp. **orkneyensis** Byt.–Salz (Orkney Is)..
Ingrailed Clay

2121 **dahlii** Hb. ....
Barred Chestnut

2122 **brunnea** D. & S. ....
Purple Clay

2123 **rubi** View. ....
Small Square-spot

2124 **florida** Schmidt ....
Fen Square-spot

XESTIA Hb.

2125 **alpicola** Zett.
ssp. **alpina** Humph. & Westw. ....
Northern Dart

2126 **c-nigrum** Linn. ....
Setaceous Hebrew Character

2127 **ditrapezium** D. & S. ....
Triple-spotted Clay

2128 **triangulum** Hufn. ....
Double Square-spot

2129 **ashworthii** Doubl. ....
Ashworth's Rustic

2130 **baja** D. & S. ....
Dotted Clay

2131 **rhomboidea** Esp. ....
Square-spotted Clay

2132 **castanea** Esp. ....
Neglected Rustic

2133 **sexstrigata** Haw. ....
Six-striped Rustic

2134 **xanthographa** D. & S. ....
Square-spot Rustic

2135 **agathina** Dup.
ssp. **agathina** Dup. ....
Heath Rustic

agathina Dup.
ssp. **hebridicola** Stdgr (Hebrides)..
Heath Rustic

NAENIA Steph.

2136 **typica** Linn. ....
The Gothic

EUROIS Hb.

2137 **occulta** Linn. ....
Great Brocade

ANAPLECTOIDES McDunn.

2138 **prasina** D. & S. ....
Green Arches

CERASTIS Ochs.

2139 **rubricosa** D. & S. ....
Red Chestnut

2140 **leucographa** D. & S. ....
White-marked

MESOGONA Boisd.

*2141 **acetosellae** D. & S. ....
Pale Stigma

HADENINAE

ANARTA Ochs.

2142 **myrtilli** Linn. ....
Beautiful Yellow Underwing

2143 **cordigera** Thunb. ....
Small Dark Yellow Underwing

2144 **melanopa** Thunb. ....
Broad-bordered White
Underwing

DISCESTRA Hamps.

2145 **trifolii** Hufn. ....
The Nutmeg

LACINIPOLIA McDunn.

*2146 **renigera** Steph. ....
Kidney-spotted Minor

HADA Billb.

2147 **nana** Hufn. ....
The Shears

POLIA Ochs.

2148 **bombycina** Hufn. ....
Pale Shining Brown

2149 **hepatica** Cl. ....
Silvery Arches

2150 **nebulosa** Hufn. ....
Grey Arches

PACHETRA Guen.

2151 **sagittigera** Hufn.
ssp. **britannica** Turn. ....
Feathered Ear

SIDERIDIS Hb.

2152 **albicolon** Hb. ....
White Colon

HELIOPHOBUS Boisd.

2153 **reticulata** Goeze
ssp. **marginosa** Haw. ....
Bordered Gothic

reticulata Goeze
ssp. **hibernica** Cock. (Ireland)..
Bordered Gothic

MAMESTRA Ochs.

2154 **brassicae** Linn. ....
Cabbage Moth

MELANCHRA Hb.

2155 **persicariae** Linn. ....
Dot Moth

LACANOBIA Billb.

2156 **contigua** D. & S. ....
Beautiful Brocade

2157 **w-latinum** Hufn. ....
Light Brocade

2158 **thalassina** Hufn. ....
Pale-shouldered Brocade

2159 **suasa** D. & S. ....
Dog's Tooth

2160 **oleracea** Linn. ....
Bright-line Brown-eye

*2161 **blenna** Hb. ....
The Stranger

PAPESTRA Sukh.

2162 **biren** Goeze ....
Glaucous Shears

CERAMICA Guen.

2163 **pisi** Linn. ....
Broom Moth

HECATERA Guen.

2164 **bicolorata** Hufn. ....
Broad-barred White

2165 **dysodea** D. & S. ....
Small Ranunculus

HADENA Schr.

2166 **rivularis** Fabr. ....
The Campion

2167 **perplexa** D. & S.
ssp. **perplexa** D. & S. ....
Tawny Shears

perplexa D. & S.
ssp. **capsophila** Dup. ....
Pod Lover

2168 **irregularis** Hufn. ....
Viper's Bugloss

2169 **luteago** D. & S.
ssp. **barrettii** Doubl. ....
Barrett's Marbled Coronet

2170 **compta** D. & S. ....
Varied Coronet

2171 **confusa** Hufn. ....
Marbled Coronet

2172 **albimacula** Borkh. ....
White Spot

2173 **bicruris** Hufn. ....
The Lychnis

2174 **caesia** D. & S.
ssp. **mananii** Gregs. ....
The Grey

ERIOPYGODES Hamps.

2175 **imbecilia** Fabr. ....
The Silurian

CERAPTERYX Curt.

2176 **graminis** Linn. ....
Antler Moth

THOLERA Hb.

2177 **cespitis** D. & S. ....
Hedge Rustic

2178 **decimalis** Poda ....
Feathered Gothic

PANOLIS Hb.

2179 **flammea** D. & S. ....
Pine Beauty

BRITHYS Hb.

*2180 **crini** Fabr.
ssp. **pancratii** Cyr. ....
Kew Arches

EGIRA Dup.

2181 **conspicillaris** Linn. ....
Silver Cloud

ORTHOSIA Ochs.

2182 **cruda** D. & S. ....
Small Quaker

2183 **miniosa** D. & S. ....
Blossom Underwing

2184 **opima** Hb. ....
Northern Drab

2185 **populeti** Fabr. ....
Lead-coloured Drab

2186 **gracilis** D. & S. ....
Powdered Quaker

2187 **stabilis** D. & S. ....
Common Quaker

2188 **incerta** Hufn. ....
Clouded Drab

2189 **munda** D. & S. ....
Twin-spotted Quaker

2190 **gothica** Linn. ....
Hebrew Character

MYTHIMNA Ochs.

2191 **turca** Linn. ....
Double Line

2192 **conigera** D. & S. ....
Brown-line Bright-eye

2193 **ferrago** Fabr. ....
The Clay

2194 **albipuncta** D. & S. ....
White-point

2195 **vitellina** Hb. ....
The Delicate

2196 **pudorina** D. & S. ....
Striped Wainscot

2197 **straminea** Treit. ....
Southern Wainscot

2198 **impura** Hb.
ssp. **impura** Hb. ....
Smoky Wainscot

impura Hb.
ssp. **scotica** Cock. (Scotland) ....
Smoky Wainscot

2199 **pallens** Linn. ....
Common Wainscot

2200 **favicolor** Barr. ....
Mathew's Wainscot

2201 **litoralis** Curt. ....
Shore Wainscot

2202 **l-album** Linn. ....
L-album Wainscot

2203 **unipuncta** Haw. ....
White-speck

2204 **obsoleta** Hb. ....
Obscure Wainscot

2205 **comma** Hb. ....
Shoulder-striped Wainscot

2206 **putrescens** Hb. ....
Devonshire Wainscot

*2207 **commoides** Guen. ....

2208 **loreyi** Dup. ....
The Cosmopolitan

SENTA Steph.

2209 **flammea** Curt. ....
Flame Wainscot

GRAPHANIA Hamps.

*2210 **dives** Philp. ....
The Maori

CUCULLIINAE

CUCULLIA Schr.

2211 **absinthii** Linn. ....
The Wormwood

*2212 **argentea** Hufn. ....
Green Silver-spangled Shark

*2213 **artemisiae** Hufn. ....
Scarce Wormwood

2214 **chamomillae** D. & S. ....
Chamomile Shark

*2215 **lactucae** D. & S. ....
Lettuce Shark

2216 **umbratica** Linn. ....
The Shark

2217 **asteris** D. & S. ....
Star-wort

2218 **gnaphalii** Hb.
ssp. **occidentalis** Bours. ....
The Cudweed

2219 **lychnitis** Ramb. ....
Striped Lychnis

*2220 **scrophulariae** D. & S. ....
Water Betony

2221 **verbasci** Linn. ....
The Mullein

*2222 **prenanthis** Boisd. ....
False Water Betony

CALOPHASIA Steph.

2223 **lunula** Hufn. ....
Toadflax Brocade

*2224 **platyptera** Esp. ....
Antirrhinum Brocade

BRACHYLOMIA Hamps.

2225 **viminalis** Fabr. ....
Minor Shoulder-knot

LEUCOCHLAENA Hamps.

2226 **oditis** Hb. ....
Beautiful Gothic

BRACHIONYCHA Hb.

2227 **sphinx** Hufn. ....
The Sprawler

2228 **nubeculosa** Esp. ....
Rannoch Sprawler

DASYPOLIA Guen.

2229 **templi** Thunb. ....
Brindled Ochre

APOROPHYLA Guen.

2230 **australis** Boisd.
ssp. **pascuea** Humph. & Westw. ....
Feathered Brindle

2231 **lutulenta** D. & S.
ssp. **lutulenta** D. & S. ....
Deep-brown Dart

lutulenta D. & S.
ssp. **lueneburgensis** Freyer ....
Northern Deep-brown Dart

2232 **nigra** Haw. ....
Black Rustic

LITHOMOIA Hb.

2233 **solidaginis** Hb. ....
Golden-rod Brindle

SCOTOCHROSTA Led.

*2234 **pulla** D. & S. ....
Ash Shoulder-knot

LITHOPHANE Hb.

2235 **semibrunnea** Haw. ....
Tawny Pinion

2236 **socia** Hufn. ....
Pale Pinion

2237 **ornitopus** Hufn.
ssp. **lactipennis** Dadd ....
Grey Shoulder-knot

2238 **furcifera** Hufn.
ssp. **furcifera** Hufn. ....
The Conformist

furcifera Hufn.
ssp. **suffusa** Tutt (Wales) ....
The Conformist

2239 **lamda** Fabr. ....
The Nonconformist

2240 **leautieri** Boisd.
ssp. **hesperica** Bours. ....
Blair's Shoulder-knot

XYLENA Ochs.

2241 **vetusta** Hb. ....
Red Sword-grass

2242 **exsoleta** Linn. ....
Sword-grass

XYLOCAMPA Guen.

2243 **areola** Esp. ....
Early Grey

MEGANEPHRIA Hb.

*2244 **bimaculosa** Linn. ....
Double-spot Brocade

ALLOPHYES Tams

2245 **oxyacanthae** Linn. ....
Green-brindled Crescent

VALERIA Steph.

*2246 **oleagina** D. & S. ....
Green-brindled Dot

DICHONIA Hb.

2247 **aprilina** Linn. ....
Merveille du Jour

DRYOBOTODES Warr.

2248 **eremita** Fabr. ....
Brindled Green

BLEPHARITA Hamps.

2249 **satura** D. & S. ....
Beautiful Arches

2250 **adusta** Esp. ....
Dark Brocade

TRIGONOPHORA Hb.

2251 **flammea** Esp. ....
Flame Brocade

POLYMIXIS Hb.

2252 **flavicincta** D. & S. ....
Large Ranunculus

2253 **xanthomista** Hb.
ssp. **statices** Gregs. ....
Black-banded

ANTITYPE Hb.

2254 **chi** Linn. ....
Grey Chi

EUMICHTIS Hb.

2255 **lichenea** Hb.
ssp. **lichenea** Hb. ....
Feathered Ranunculus

lichenea Hb.
ssp. **scillonea** Rich. (Scilly Is) ....
Feathered Ranunculus

EUPSILIA Hb.

2256 **transversa** Hufn. ....
The Satellite

JODIA Hb.

2257 **croceago** D. & S. ....
Orange Upperwing

CONISTRA Hb.

2258 **vaccinii** Linn. ....
The Chestnut

2259 **ligula** Esp. ....
Dark Chestnut

2260 **rubiginea** D. & S. ....
Dotted Chestnut

2261 **erythrocephala** D. & S. ....
Red-headed Chestnut

AGROCHOLA Hb.

2262 **circellaris** Hufn. ....
The Brick

2263 **lota** Cl. ....
Red-line Quaker

2264 **macilenta** Hb. ....
Yellow-line Quaker

2265 **helvola** Linn. ....
Flounced Chestnut

2266 **litura** Linn. ....
Brown-spot Pinion

2267 **lychnidis** D. & S. ....
Beaded Chestnut

PARASTICHTIS Hb.

2268 **suspecta** Hb. ....
The Suspected

ATETHMIA Hb.

2269 **centrago** Haw. ....
Centre-barred Sallow

OMPHALOSCELIS Hamps.

2270 **lunosa** Haw. ....
Lunar Underwing

XANTHIA Ochs.

2271 **citrago** Linn. ....
Orange Sallow

2272 **aurago** D. & S. ....
Barred Sallow

2273 **togata** Esp. ....
Pink-barred Sallow

2274 **icteritia** Hufn. ....
The Sallow

2275 **gilvago** D. & S. ....
Dusky-lemon Sallow

2276 **ocellaris** Borkh. ....
Pale-lemon Sallow

ACRONICTINAE

MOMA Hb.

2277 **alpium** Osb. ....
Scarce Merveille du Jour

ACRONICTA Ochs.

2278 **megacephala** D. & S. ....
Poplar Grey

2279 **aceris** Linn. ....
The Sycamore

2280 **leporina** Linn. ....
The Miller

2281 **alni** Linn. ....
Alder Moth

*2282 **cuspis** Hb. ....
Large Dagger

2283 **tridens** D. & S. ....
Dark Dagger

2284 **psi** Linn. ....
Grey Dagger

2285 **strigosa** D. & S. ....
Marsh Dagger

2286 **menyanthidis** Esp.
ssp. **menyanthidis** Esp. ....
Light Knot Grass

menyanthidis Esp.
ssp. **scotica** Tutt (Scotland) ....
Light Knot Grass

2287 **auricoma** D. & S. ....
Scarce Dagger

2288 **euphorbiae** D. & S.
ssp. **myricae** Guen. ....
Sweet Gale Moth

2289 **rumicis** Linn. ....
Knot Grass

SIMYRA Ochs.

2290 **albovenosa** Goeze ....
Reed Dagger

CRANIOPHORA Snell.

2291 **ligustri** D. & S. ....
The Coronet

CRYPHIA Hb.

*2292 **algae** Fabr. ....
Tree-lichen Beauty

2293 **domestica** Hufn. ....
Marbled Beauty

*2294 **raptricula** D. & S. ....
Marbled Grey

2295 **muralis** Forst.
ssp. **muralis** Forst. ....
Marbled Green

muralis Forst.
ssp. **impar** Warr. (Cambridge) ....
Marbled Green

muralis Forst.
ssp. **westroppi** Cock. (Ireland: Co. Cork)..........
Marbled Green

AMPHIPYRINAE

TATHORHYNCHUS Led.

*2296 **exsiccata** Led. ....
Levant Blackneck

AMPHIPYRA Ochs.

2297 **pyramidea** Linn. ....
Copper Underwing

2298 **berbera** Rungs
ssp. **svenssoni** Fletch. ....
Svensson's Copper Underwing

2299 **tragopogonis** Cl. ....
Mouse Moth

MORMO Ochs.

2300 **maura** Linn. ....
Old Lady

DYPTERYGIA Steph.

2301 **scabriuscula** Linn. ....
Bird's Wing

RUSINA Steph.

2302 **ferruginea** Esp. ....
Brown Rustic

THALPOPHILA Hb.

2303 **matura** Hufn. ....
Straw Underwing

TRACHEA Ochs.

2304 **atriplicis** Linn. ....
Orache Moth

EUPLEXIA Steph.

2305 **lucipara** Linn. ....
Small Angle Shades

PHLOGOPHORA Treit.

2306 **meticulosa** Linn. ....
Angle Shades

PSEUDENARGIA Bours.

*2307 **ulicis** Stdgr ....
The Berber

CALLOPISTRIA Hb.

*2308 **juventina** Stoll ....
The Latin

*2309 **latreillei** Dup. ....

EUCARTA Led.

*2310 **amethystina** Hb. ....
Cumberland Gem

IPIMORPHA Hb.

2311 **retusa** Linn. ....
Double Kidney

2312 **subtusa** D. & S. ....
The Olive

ENARGIA Hb.

2313 **paleacea** Esp. ....
Angle-striped Sallow

2314 **ypsillon** D. & S. ....
Dingy Shears

DICYCLA Guen.

2315 **oo** Linn. ....
Heart Moth

COSMIA Ochs.

2316 **affinis** Linn. ....
Lesser-spotted Pinion

2317 **diffinis** Linn. ....
White-spotted Pinion

2318 **trapezina** Linn. ....
The Dun-bar

2319 **pyralina** D. & S. ....
Lunar-spotted Pinion

HYPPA Dup.

2320 **rectilinea** Esp. ....
The Saxon

APAMEA Ochs.

2321 **monoglypha** Hufn. ....
Dark Arches

2322 **lithoxylaea** D. & S.
Light Arches ....

2323 **sublustris** Esp.
Reddish Light Arches ....

2324 **exulis** Lefeb.
ssp. **exulis** Lefeb.
The Exile ....

exulis Lefeb.
ssp. **assimilis** Doubl.
Northern Arches ....

2325 **oblonga** Haw.
Crescent Striped ....

2326 **crenata** Hufn.
Clouded-bordered Brindle ....

2327 **epomidion** Haw.
Clouded Brindle ....

*2328 **lateritia** Hufn.
Scarce Brindle ....

2329 **furva** D. & S.
ssp. **britannica** Cock.
The Confused ....

2330 **remissa** Hb.
Dusky Brocade ....

2331 **unanimis** Hb.
Small Clouded Brindle ....

2332 **pabulatricula** Brahm
Union Rustic ....

2333 **anceps** D. & S.
Large Nutmeg ....

2334 **sordens** Hufn.
Rustic Shoulder-knot ....

2335 **scolopacina** Esp.
Slender Brindle ....

2336 **ophiogramma** Esp.
Double Lobed ....

OLIGIA Hb.

2337 **strigilis** Linn.
Marbled Minor ....

2338 **versicolor** Borkh.
Rufous Minor ....

2339 **latruncula** D. & S.
Tawny Marbled Minor ....

2340 **fasciuncula** Haw.
Middle-barred Minor ....

MESOLIGIA Bours.

2341 **furuncula** D. & S. ....
Cloaked Minor

2342 **literosa** Haw. ....
Rosy Minor

MESAPAMEA Hein.

2343 **secalis** Linn. ....
Common Rustic

PHOTEDES Led.

2344 **captiuncula** Treit.
ssp. **expolita** Staint. ....
Least Minor

captiuncula Treit.
ssp. **tincta** Kane (Ireland)..
Least Minor

2345 **minima** Haw. ....
Small Dotted Buff

2346 **morrisii** Dale
ssp. **morrisii** Dale ....
Morris's Wainscot

morrisii Dale
ssp. **bondii** Knaggs ....
Bond's Wainscot

2347 **extrema** Hb. ....
The Concolorous

2348 **elymi** Treit. ....
Lyme Grass

2349 **fluxa** Hb. ....
Mere Wainscot

2350 **pygmina** Haw. ....
Small Wainscot

2351 **brevilinea** Fenn ....
Fenn's Wainscot

EREMOBIA Steph.

2352 **ochroleuca** D. & S. ....
Dusky Sallow

LUPERINA Boisd.

2353 **testacea** D. & S. ....
Flounced Rustic

2354 **nickerlii** Freyer
ssp. **gueneei** Doubl. ....
Sandhill Rustic

nickerlii Freyer
ssp. **leechi** Goater (Cornwall) ....
Sandhill Rustic

nickerlii Freyer
ssp. **knilli** Bours. (S.W. Ireland) ....
Sandhill Rustic

2355 **dumerilii** Dup. ....
Dumeril's Rustic

2356 **zollikoferi** Freyer ....
Scarce Arches

AMPHIPOEA Billb.

2357 **lucens** Freyer ....
Large Ear

2358 **fucosa** Freyer
ssp. **paludis** Tutt ....
Saltern Ear

2359 **crinanensis** Burr. ....
Crinan Ear

2360 **oculea** Linn. ....
Ear Moth

HYDRAECIA Guen.

2361 **micacea** Esp. ....
Rosy Rustic

2362 **petasitis** Doubl. ....
The Butterbur

2363 **osseola** Stdgr
ssp. **hucherardi** Mab. ....
Marsh Mallow Moth

GORTYNA Ochs.

2364 **flavago** D. & S. ....
Frosted Orange

2365 **borelii** Pierr.
ssp. **lunata** Freyer ....
Fisher's Estuarine Moth

CALAMIA Hb.

2366 **tridens** Hufn.
ssp. **occidentalis** Cock. ....
Burren Green

CELAENA Steph.

2367 **haworthii** Curt. ....
Haworth's Minor

2368 **leucostigma** Hb.
ssp. **leucostigma** Hb. ....
The Crescent

leucostigma Hb.
ssp. **scotica** Cock. (Scotland) ....
The Crescent

NONAGRIA Ochs.

2369 **typhae** Thunb. ....
Bulrush Wainscot

ARCHANARA Walk.

2370 **geminipuncta** Haw. ....
Twin-spotted Wainscot

2371 **dissoluta** Treit. ....
Brown-veined Wainscot

2372 **neurica** Hb. ....
White-mantled Wainscot

2373 **sparganii** Esp. ....
Webb's Wainscot

2374 **algae** Esp. ....
Rush Wainscot

RHIZEDRA Warr.

2375 **lutosa** Hb. ....
Large Wainscot

SEDINA Urb.

2376 **buettneri** Her. ....
Blair's Wainscot

ARENOSTOLA Hamps.

2377 **phragmitidis** Hb. ....
Fen Wainscot

ORIA Hb.

2378 **musculosa** Hb. ....
Brighton Wainscot

COENOBIA Steph.

2379 **rufa** Haw. ....
Small Rufous

CHARANYCA Billb.

2380 **trigrammica** Hufn. ....
Treble Lines

HOPLODRINA Bours.

2381 **alsines** Brahm ....
The Uncertain

2382 **blanda** D. & S. ....
The Rustic

*2383 **superstes** Ochs. ....
Powdered Rustic

2384 **ambigua** D. & S. ....
Vine's Rustic

SPODOPTERA Guen.

2385 **exigua** Hb. ....
Small Mottled Willow

2386 **littoralis** Boisd. ....
Mediterranean Brocade

CARADRINA Ochs.

2387 **morpheus** Hufn. ....
Mottled Rustic

*2388 **flavirena** Guen. ....

clavipalpis Scop. ....
Pale Mottled Willow

PERIGEA Guen.

*2390 **conducta** Walk. ....
The African

CHILODES H.-S.

2391 **maritimus** Tausch. ....
Silky Wainscot

ATHETIS Hb.

2392 **pallustris** Hb. ....
Marsh Moth

ACOSMETIA Steph.

2393 **caliginosa** Hb. ....
Reddish Buff

STILBIA Steph.

2394 **anomala** Haw. ....
The Anomalous

SYNTHYMIA Hb.

*2395 **fixa** Fabr. ....
The Goldwing

ELAPHRIA Hb.

2396 **venustula** Hb. ....
Rosy Marbled

PANEMERIA Hb.

2397 **tenebrata** Scop. ....
Small Yellow Underwing

HELIOTHINAE

PERIPHANES Hb.

2398 **delphinii** Linn. ....
Pease Blossom

PYRRHIA Hb.

2399 **umbra** Hufn. ....
Bordered Sallow

HELICOVERPA Hard.

2400 **armigera** Hb. ....
Scarce Bordered Straw

HELIOTHIS Ochs.

2401 **viriplaca** Hufn. ....
Marbled Clover

2402 **maritima** Grasl.
ssp. **warneckei** Bours. ....
Shoulder-striped Clover

* **maritima** Grasl.
ssp. **bulgarica** Draudt (S.E. Europe) ....
Shoulder-striped Clover

2403 **peltigera** D. & S. ....
Bordered Straw

*2404 **nubigera** H.–S. ....
Eastern Bordered Straw

PROTOSCHINIA Hard.

2405 **scutosa** D. & S. ....
Spotted Clover Moth

SCHINIA Hb.

*2406 **rivulosa** Guen. ....
Scarce Meal-moth

ACONTIINAE

EUBLEMMA Hb.

2407 **ostrina** Hb. ....
Purple Marbled

2408 **parva** Hb. ....
Small Marbled

*2409 **noctualis** Hb. ....
Scarce Marbled

LITHACODIA Hb.

2410 **pygarga** Hufn. ....
Marbled White Spot

2411 **deceptoria** Scop. ....
Pretty Marbled

EUSTROTIA Hb.

2412 **uncula** Cl. ....
Silver Hook

DELTOTE R.L.

2413 **bankiana** Fabr. ....
Silver Barred

EMMELIA Hb.

2414 **trabealis** Scop. ....
Spotted Sulphur

ACONTIA Ochs.

2415 **lucida** Hufn. ....
Pale Shoulder

*2416 **aprica** Hb. ....
The Nun

*2417 **nitidula** Fabr. ....
Brixton Beauty

CHLOEPHORINAE

EARIAS Hb.

2418 **clorana** Linn. ....
Cream-bordered Green Pea

*2419 **biplaga** Walk. ....
Spiny Bollworm

*2420 **insulana** Boisd. ....
Egyptian Bollworm

BENA Billb.

2421 **prasinana** Linn. ....
Scarce Silver-lines

PSEUDOIPS Hb.

2422 **fagana** Fabr.
ssp. **britannica** Warr. ....
Green Silver-lines

SARROTHRIPINAE

NYCTEOLA Hb.

2423 **revayana** Scop. ....
Oak Nycteoline

*2424 **degenerana** Hb. ....
Sallow Nycteoline

PANTHEINAE

COLOCASIA Ochs.

2425 **coryli** Linn. ....
Nut-tree Tussock

CHARADRA Walk.

*2426 **deridens** Guen. ....
Marbled Tuffet

RAPHIA Hb.

*2427 **frater** Grote ....
The Brother

PLUSIINAE

CHRYSODEIXIS Hb.

*2428 **chalcites** Esp. ....
Golden Twin-spot

*2429 **acuta** Walk. ....
Tunbridge Wells Gem

CTENOPLUSIA Dufay

*2430 **limbirena** Guen. ....
Scar Bank Gem

*2431 **accentifera** Lefeb. ....

TRICHOPLUSIA McDunn.

2432 **ni** Hb. ....
The Ni Moth

DIACHRYSIA Hb.

2433 **orichalcea** Fabr. ....
Slender Burnished Brass

2434 **chrysitis** Linn. ....
Burnished Brass

2435 **chryson** Hb. ....
Scarce Burnished Brass

MACDUNNOUGHIA Kostr.

2436 **confusa** Steph. ....
Dewick's Plusia

POLYCHRYSIA Hb.

2437 **moneta** Fabr. ....
Golden Plusia

EUCHALCIA Hb.

*2438 **variabilis** Pill. ....
Purple-shaded Gem

PLUSIA Ochs.

2439 **festucae** Linn. ....
Gold Spot

2440 **putnami** Grote
ssp. **gracilis** Lempke ....
Lempke's Gold Spot

AUTOGRAPHA Hb.

2441 **gamma** Linn. ....
Silver Y

2442 **pulchrina** Haw. ....
Beautiful Golden Y

2443 **jota** Linn. ....
Plain Golden Y

2444 **bractea** D. & S. ....
Gold Spangle

*2445 **biloba** Steph. ....
Stephens' Gem

*2446 **bimaculata** Steph. ....
Double-spotted Spangle

SYNGRAPHA Hb.

2447 **interrogationis** Linn. ....
Scarce Silver Y

*2448 **circumflexa** Linn. ....
Yorkshire Y

ABROSTOLA Ochs.

2449 **trigemina** Werneb. ....
Dark Spectacle

2450 **triplasia** Linn. ....
The Spectacle

CATOCALINAE

CATOCALA Schr.

2451 **fraxini** Linn. ....
Clifden Nonpareil

2452 **nupta** Linn. ....
Red Underwing

*2453 **electa** View. ....
Rosy Underwing

2454 **promissa** D. & S. ....
Light Crimson Underwing

2455 **sponsa** Linn. ....
Dark Crimson Underwing

MINUCIA Moore

2456 **lunaris** D. & S. ....
Lunar Double-stripe

CLYTIE Hb.

*2457 **illunaris** Hb. ....

CAENURGINA McDunn.

*2458 **crassiuscula** Haw. ....
The Double-barred

MOCIS Hb.

*2459 **trifasciata** Steph. ....
The Triple-barred

DYSGONIA Hb.

*2460 **algira** Linn. ....

GRAMMODES Guen.

*2461 **stolida** Fabr. ....
The Geometrician

CALLISTEGE Hb.

2462 **mi** Cl. ....
Mother Shipton

EUCLIDIA Ochs.

2463 **glyphica** Linn. ....
Burnet Companion

OPHIDERINAE

CATEPHIA Ochs.

2464 **alchymista** D. & S. ....
The Alchymist

TYTA Billb.

2465 **luctuosa** D. & S. ....
The Four-spotted

LYGEPHILA Billb.

2466 **pastinum** Treit. ....
The Blackneck

2467 **craccae** D. & S. ....
Scarce Blackneck

SYNEDOIDA Edw.

*2468 **grandirena** Haw. ....
Great Kidney

SCOLIOPTERYX Germ.

2469 **libatrix** Linn. ....
The Herald

PHYTOMETRA Haw.

2470 **viridaria** Cl. ....
Small Purple-barred

ANOMIS Hb.

*2471 **sabulifera** Guen. ....

COLOBOCHYLA Hb.

2472 **salicalis** D. & S. ....
Lesser Belle

LASPEYRIA Germ.

2473 **flexula** D. & S. ....
Beautiful Hook-tip

RIVULA Guen.

2474 **sericealis** Scop. ....
Straw Dot

PARASCOTIA Hb.

2475 **fuliginaria** Linn. ....
Waved Black

HYPENINAE

HYPENA Schr.

2476 **crassalis** Fabr. ....
Beautiful Snout

2477 **proboscidalis** Linn. ....
The Snout

2478 **obsitalis** Hb. ....
Bloxworth Snout

*2479 **obesalis** Treit. ....
Paignton Snout

2480 **rostralis** Linn. ....
Buttoned Snout

PLATHYPENA Grote

*2481 **scabra** Fabr. ....
Black Snout

SCHRANKIA Hb.

2482 **taenialis** Hb. ....
White-line Snout

2483 **intermedialis** Reid ....
Autumnal Snout

2484 **costaestrigalis** Steph. ....
Pinion-streaked Snout

HYPENODES Doubl.

2485 **turfosalis** Wocke ....
Marsh Oblique-barred

EPIZEUXIS Hb.

*2486 **aemula** Hb. ....
Waved Tabby

*2487 **lubricalis** Geyer ....
Twin-striped Tabby

HERMINIA Lat.

2488 **strigilata** Linn. ....
Common Fan-foot

2489 **tarsipennalis** Treit. ....
The Fan-foot

*2490 **lunalis** Scop. ....

2491 **tarsicrinalis** Knoch ....

2492 **nemoralis** Fabr. ....
Small Fan-foot

MACROCHILO Hb.

2493 **cribrumalis** Hb. ....
Dotted Fan-foot

PARACOLAX Hb.

2494 **derivalis** Hb. ....
Clay Fan-foot

TRISATELES Tams

2495 **emortualis** D. & S. ....
Olive Crescent

Addenda & Corrigenda

242 **translucens** Meyr. ....
metonella Pier. & Metc.

243 **dubiella** Stt. ....
turicensis Müll.-Rutz

*247b **murariella** Stdgr ....